ENGINEERING DESIGN:
A MODERN APPROACH

Roy Edwards
161 Millspaugh Dr.
Ffld, CT 06430
254 - 8015

ENGINEERING DESIGN:
A MODERN APPROACH

B. S. DHILLON
PROFESSOR AND CHAIRMAN
UNIVERSITY OF OTTAWA

IRWIN

Chicago • Bogotá • Boston • Buenos Aires • Caracas
London • Madrid • Mexico City • Sydney • Toronto

**This book is affectionately dedicated to my late uncle
Piara S. Mann.**

Irwin Book Team
Publisher: *Tom Casson*
Senior sponsoring editor: *Elizabeth A. Jones*
Senior developmental editor: *Kelley Butcher*
Senior marketing manager: *Brian Kibby*
Project editor: *Margaret Rathke*
Production supervisor: *Dina L. Treadaway*
Manager, Prepress Purchasing: *Kim Meriwether*
Senior Designer: *Heidi J. Baughman*
Coordinator, graphics, & desktop services: *Keri Kunst*
Photo researcher: *Charlotte Goldman*
Compositor: *Interactive Composition Corporation*
Typeface: *10/12 Times Roman*
Printer: *R. R. Donnelley & Sons Company*

Times Mirror
Higher Education Group

Library of Congress Cataloging-in-Publication Data

Dhillon, B. S.
    Engineering design: a modern approach/B. S. Dhillon.
      p. cm.
    Includes index.
    ISBN 0–256–18312–0
    1. Engineering design. I. Title.
  TA174.D42 1996
  620'.0042–dc20   95–37446

Printed in the United States of America
1 2 3 4 5 6 7 8 9 0 DOC 2 1 0 9 8 7 6 5

AUTHOR BIOGRAPHY

Dr. B. S. Dhillon is a full Professor and Chairman of Mechanical Engineering and Director of the Engineering Management Program at the University of Ottawa. He has published over 235 articles on reliability engineering and related areas. He is or has been on the editorial boards of several journals. In addition, he has written 15 books on various aspects of system reliability, safety, human factors, maintainability, and engineering management, published by Wiley (1981), Van Nostrand (1982), Butterworth (1983), Marcel Dekker (1984), Pergamon (1986), etc. His books on reliability have been translated into several languages, including Russian, Chinese, and German. He served as General Chairman of two international conferences on reliability and quality control, held in Los Angeles and Paris in 1987.

Dr. Dhillon is a recipient of the American Society for Quality Control Austin J. Bonis Reliability Award, the Society of Reliability Engineers' Merit Award, the Gold Medal of Honor (American Biographical Institute), and Faculty of Engineering Galinski Award for Excellence in Reliability Engineering Research. He is a Registered Professional Engineer in Ontario and is listed in American Men and Women of Science, Men of Achievements, International Dictionary of Biography, Who's Who in International Intellectuals, and Who's Who in Technology.

Dr. Dhillon has been teaching engineering design for many years and attended the University of Wales, where he received a B.S. in electrical and electronic engineering and an M.S. in mechanical engineering. He received a Ph.D. in industrial engineering from the University of Windsor.

PREFACE

People have been doing engineering designs of objects and structures for thousands of years: the Egyptian pyramids, the great wall of China, etc. Since those times, considerable progress has been made in engineering designs.

Today, traditional design approaches are no longer effective, as products have become more complex and sophisticated, consumer driven, government controlled, etc. This has led to changes in the teaching materials for college-level students, and graduates are now armed with the knowledge to vigorously meet the new challenges of the industrial sector. In many engineering schools, students are taught not only traditional design approaches, but also such subjects as human factors, total quality management, reliability, maintainability, computer-aided design, concurrent engineering, value analysis, and life-cycle costing.

The author has been teaching design-oriented courses for many years and has successfully tailored modern concepts into them. This text is the result of the experience thus gained and the feedback received from the graduates. The book is intended to fulfill the stated vital need and is written in such a manner that the reader needs no previous knowledge to understand its contents. In general, the book focuses more on the structure of the concepts than on the minute details. Over 600 references to relevant literature are provided for the reader who wishes to delve more deeply into particular topics or the general design process. In addition, the book contains many examples and their corresponding solutions, along with ten case studies and challenging problems to test comprehension.

The book is composed of 14 chapters plus an Appendix. The first chapter reviews technical writing, since the design documentation is as important as the design itself. Skills in technical writing are absolutely essential for the success of a new design.

Chapter 2 reviews the basic engineering design areas. The design process is covered in Chapter 3. Chapter 4 presents important design tools: optimization methods, project management approaches, creativity techniques, marketing approaches and procedures, and computer-aided design (CAD). Essential probability and statistics concepts are covered in Chapter 5, including such topics as: probability distributions, statistical tests, confidence limits, and linear regression analyses.

Chapter 6 describes quality assurance principles, including: Taguchi methods, quality control charts, and total quality management. Chapters 7 to 10 discuss reliability, maintainability, safety, and human factor considerations, respectively. Economic analysis and cost considerations are covered in Chapter 11, including depreciation methods, life-cycle costing, and cost estimation models. Chapter 12 addresses three areas: material selection, manufacturing, and environmental considerations. Value analysis, configuration management, concurrent engineering, and reverse engineering, all of which directly or indirectly impact engineering design, are covered in Chapter 13. Finally, Chapter 14 presents ethics and legal considerations, including codes of ethics, liability, warranties, copyrights, and patents. The Appendix lists 301 references on engineering design and related areas.

The basic objective of this text is to help in teaching engineering design to junior and senior undergraduates. As the book contains most of the topics related to engineering design, this should allow course instructors to tailor their coverage as required. Engineering professionals and others involved with design should also find it a useful document.

The author sincerely wishes to thank the editorial people at Richard D. Irwin, Inc., for their close interest in this project. He is also indebted to referees and other people for their invisible input, as well as to the organizations that permitted the use of their images. The author also wishes to thank the following reviewers for their contribution to the content and quality of this book: Robert M. Briber, University of Maryland; R. C. Yalamanchili, University of Pittsburgh; Gladius Lewis, University of Memphis; Edward H. Lin, University of Memphis; James C. Shahan, Iowa State University; and Massoud S. Tavakoli, GMI Engineering and Management Institute. Last, but not the least, I thank my wife, Rosy, for typing the original manuscript, her help in proofreading, and the coffee-making!

B. S. Dhillon
University of Ottawa

CONTENTS

Chapter 10

HUMAN FACTORS CONSIDERATIONS   186

Chapter 11

ECONOMIC ANALYSIS AND COST ESTIMATION   202

Chapter 12

MATERIALS SELECTION, MANUFAC-TURING, AND ENVIRONMENTAL DESIGN   220

Chapter 13

VALUE ENGINEERING, CONFIGURATION MANAGEMENT, CONCURRENT ENGINEERING, AND REVERSE ENGINEERING 242

Chapter 14

ETHICAL AND LEGAL FACTORS 254

Appendix

BIBLIOGRAPHY: ENGINEERING DESIGN LITERATURE 263

1

TECHNICAL WRITING

1.1 INTRODUCTION

Generally speaking, it is essential to have good communication skills regardless of one's career path. The ability to speak and write effectively plays an important role in securing a job, getting raises and promotions, etc.

In the engineering and scientific professions, communication skills are as important as in other fields. Some years ago, 245 distinguished engineers [1,2] and 837 experienced business persons [1,3] were surveyed, asking, "How important is the writing that you do?" Over 95 percent of the engineers responded by stating that it was either critically or very important, as opposed to 74 percent of the business persons. In another survey, 595 engineering alumni of the University of California, Berkeley [4], were asked if writing skills had aided their advancement. Over 70 percent responded affirmatively, and 95 percent stated that they would certainly consider writing ability a factor in decisions concerning promotions or hiring.

According to two survey studies [2,4], roughly 25 percent of an engineering professional's time is spent on writing. It is interesting to note that typical verbal communication time varies from 20 percent (junior engineers) to 65 percent (chief executives) of the total work time.

The task of technical writing is not confined only to modern man. The ancients thousands of years ago also dealt with it somewhat, in a primitive form. For example, Babylonians, Egyptians, and Phoenicians used clay tablets and papyrus scrolls [5] for recording information on agricultural yields, astronomical changes, medical practices, and mathematics. Further, a skillful organization of technical details may be found in early (500 BC) Greek documents. Known documents from that era cover a range of material, from explanations of architecture to detailed medical manuals. During the Roman times, the work of Pliny, the Elder (AD 23–79) [6] is an important document of technical and scientific writing. In this document, Pliny's writing covers such things as recommendations for workers wearing protective masks to protect them from inhaling toxic substances to possible explanations of the construction of the Giza pyramids.

In design engineering work, technical writing skills are crucial. An excellent design may be overlooked if it is poorly documented. Therefore, it is absolutely necessary for a successful designer to have good technical and writing skills. This chapter discusses some important aspects of technical writing.

1.2   TECHNICAL WRITING CHARACTERISTICS, AND WRITING STAGES

Technical writing may be defined as the writing that deals basically with scientific and technical subjects. It has several characteristics [5]: purpose, subject matter, format, style, audience, timeliness, interpretation, need, visuals, and approach. A clear-cut understanding of each of these characteristics is essential in producing effective technical reports. There are also several stages of writing, whether it is technical or otherwise. In general, the writing process is very recursive, that is, various stages occur more than once and some occur concurrently. The stages involved in writing are as follows:

1. Exploring.
2. Planning and Organizing.
3. Drafting.
4. Revising.
5. Editing.
6. Proofreading.

During the exploring stage, the writers basically identify and investigate the theme, in addition to considering the audience. Some of the tasks involved at this stage are: reviewing available relevant literature, reviewing ideas, performing relevant "experiments," and identifying factors that contribute toward the need for a document. The degree of planning and organizing may vary from one type of technical report to another. Often, planning and exploring occur simultaneously with exploring. An important factor in planning and organizing is identifying and reviewing the constraints: scheduling, cost, inclusion or exclusion of certain contents, etc.

The drafting stage is concerned with both writing the text and preparing the visuals. In the revising stage, the draft is read and reread critically with respect to such factors as organization, structure, content, and design. The writer may add, delete, and rearrange material to produce a more effective document. Editing is generally concerned with identification and correction of such things as inconsistencies, errors, and unclear or ungrammatical phrases. A technical document may need the following types of editing:

1. **Language.** Concerned with grammatical aspects of the material presented.
2. **Format.** Concerned with the actual format of the technical document, such as page design, headings, and fonts.
3. **Substance.** Concerned with reviewing the document contents for organization and consistency.

4. **Integrity.** Concerned with matching text references to corresponding references, tables, figures, and so on.

5. **Copy clarification.** Concerned with providing proper instructions to such people as compositors and graphic artists.

6. **Policy edit.** Concerned with enforcing institutional policies.

7. **Coordination.** Concerned with the administrative aspects of publishing a document.

8. **Mechanical style.** Concerned with establishing consistency in "micro" physical elements: citations, symbols, etc.

9. **Screening.** Concerned with rectifying language and numerical mistakes.

Proofreading is an equally important stage of writing and is basically concerned with examining a document for typographical inconsistencies and errors.

1.3 INFORMATION COLLECTION SOURCES

There are various ways of obtaining information related to technical writing. The basic three sources are: interviews, direct observations, and libraries [7]. Since interviews and direct observations are self-explanatory, we will concentrate on the library-related sources. Some of the important library-related sources of information are as follows:

1. **Encyclopedias.** These are generally useful for providing the initial framework for the writing search. Three examples of general encyclopedias are: Encyclopedia Britannica, Encyclopedia Americana, and the Columbia Encyclopedia. Specialized encyclopedias on many areas of science, business, and technology are also available in many libraries. Some examples are: engineering, computer science, medicine, physics, mathematics, economics, management, astronomy, chemistry, biology, and geology.

2. **Handbooks.** These provide comprehensive and useful information on a given area. Generally, there is a handbook on each specialized technical area. Some examples are: *Composite Materials Handbook* (McGraw-Hill), *Mechanical Engineers' Handbook* (Wiley), and *Civil Engineering Handbook* (McGraw-Hill).

3. **Dictionaries.** These define words and terms. Some examples are: *Penguin Dictionary of Science* (Penguin), *Chamber's Dictionary of Science and Technology* (Macmillan), and *The International Business Dictionary & Reference* (Wiley).

4. **Periodicals.** These are useful in obtaining up-to-date information on a given area of interest.

5. **Conference proceedings.** Each year, hundreds of conferences are held around the world, on many technical areas. Their proceedings contain the papers presented, which detail the latest knowledge and research in the areas of interest.

6. **Books.** These provide information compiled by other authors and researchers in the topics of interest.

7. **Computerized databases.** These are a good source for obtaining summarized information. Examples of science and technology databases are: Compendex, Aerospace database, Inspec, and National Technical Information Service (NTIS).

8. **Publications of professional societies.** Many professional societies publish documents on recent advances or ongoing research. Some of these professional societies are: American Society for Engineering Education (ASEE), American Society of Mechanical Engineers (ASME), Institute of Electrical and Electronic Engineers (IEEE), and American Institute of Physics (AIP).

1.4   TECHNICAL DOCUMENT TYPES

In the technical arena, there are various types of documents. Some of these are as follows [8]:

1. **Academic laboratory report.** This is intended to help students understand a process or procedure, test a theory, or conduct a specific test. The components of these documents include: summary, objective, equipment setup, test method, test results, analysis or interpretation, conclusions, and attachments.

2. **Occurrence report.** This explains an event or incident, and describes how and why it occurred. The report also clearly indicates the effect(s) of the event and any relevant action initiated or taken.

3. **Investigation report.** This explains how tests were performed, the given data were examined, or an investigation was conducted, with the help of tangible evidence. Typical parts of an investigation report are: summary, background, investigation details, conclusions, and evidence.

4. **Trip report.** This presents the details of trips taken to satisfy some specific objective, such as attending a short course, workshop, or conference, or visiting a client's site to gather data. The report generally includes information on who went and where, why and when, original plan for the trip duration, the degree of success, unfinished portion of the plan, any additional accomplishments, and the problems faced (if any).

5. **Progress report.** This report keeps management and other concerned people up to date regarding the progress of projects or other topics that generally span a relatively lengthy period (that is, a few weeks to several years). Topics covered include: summary, progress, current situation, and future plans.

6. **Inspection report.** This presents the findings of an inspection of the work either in progress or completed. The major components of the report are: summary, background, facts, and outcome.

7. **Letters and memoranda** [9]. Letters are used to convey information from one person to another or to a group, generally not in the same organization. In contrast, memoranda serve the same purpose as the letters but are written from one person to

another or to a group within the same organization. General guidelines for letter and memorandum writers are as follows:

- Clearly keep in mind the purpose of the letter/memorandum.
- Thoroughly understand who the readers are.
- Follow the correct format.
- Follow the abstract (A), body (B), and conclusion (C) format.
- Apply the three C's (capture, convince, and control) strategy for persuasive messages.
- Emphasize the "you" attitude by using the reader's name in the body of the memorandum or letter, anticipating questions, and replacing the pronouns "I", "me," and "we" with "you" and "your."
- Keep the text brief, and place details (if necessary) in attachments.
- Be diplomatic.
- Perform the editing function with care.
- Respond as quickly as possible.

8. **Manuals** [5]. Manuals play a critical role in such areas as engineering equipment manufacture and operation. This calls for special care at the writing stage. The basic objective of such manuals is to provide instructions/directions. Common examples of manual uses are: installing an automatic washing machine, calibrating electronic inspection equipment, installing chips on a circuit board, and specifying incoming inspection procedures. Generally, a manual contains: title page, introduction, reader identification, theory of operation, training materials, step-by-step instructions, trouble-shooting and maintenance, and frequent users' guide (summary of steps).

9. **Proposals.** These are intended to describe work that is to be done. A proposal answers questions such as, what are the advantages of the proposed solution, what is the proposed item (that is, plan, product, service, or idea), what problem is addressed, why does the described problem require attention, how is the work to be performed, when is it to be performed, how much will it cost, and how is the effectiveness of the proposed work to be measured. A proposal may be one page, even in the form of a letter, or may be as large as several volumes.

1.5 GOOD REPORT WRITERS, AND GOOD REPORT GUIDELINES

Good writers usually produce good reports. Understanding the attributes of good report writers and good reports [10] is, therefore, essential. Some of those attributes are as follows:

1. Write as objectively as possible. Do not become emotionally involved or attached to a problem or a solution.
2. Be reasonably methodical and painstaking.
3. Record whatever is learned, and keep in mind that whatever work is performed must eventually be documented.

4. Always strive for clarity in writing, and keep in mind that the written material should be simple and straightforward.

5. Accept the fact that writing is a skill that can be learned and improved.

6. Make every effort to deliver written material on time.

There are also certain qualities associated with a good report. These qualities may vary from one report to another, depending upon the objective, audience, etc. General qualities that may be associated with a good report are as follows:

1. Delivered on due date.

2. Effectively answers readers' questions as they arise.

3. Gives a good impression at first sight.

4. Reads coherently from start to end.

5. Contains an effective summary and/or conclusions (if required).

6. Gives the general impression of authority, honest work, thoroughness, etc.

7. Is written clearly and concisely and avoids vague or superfluous phrases.

8. Provides pertinent information.

9. Possesses needed front matter to characterize the report.

10. Effectively discloses its objective and scope.

11. Contains a readily discernible plan and rationale.

12. Is designed for usage by readers with various needs (for example, "abstract only" readers, "introduction and conclusions only" readers, and "entire report" readers).

13. Is free from grammatical slips, misspelled words, and typographical errors.

Many texts on technical writing have presented certain practical guidelines, which may be classified into the following categories:

1. Concept development.

2. Report presentation features.

3. Report preparation.

4. Data presentation.

5. Avoidance items.

During the initial stage of writing a technical report, it is important to answer such questions as: types of anticipated readers (for example, engineers and salesmen); sources of information; types of material to be covered (for example, technical or nontechnical); and, the purpose of the report. The "report presentation features" should include such items as: an effective abstract or summary, brief and descriptive title, clear and concise language, clear distinction between facts and opinions, emphasis on essential points, listing and description of background information, sufficient depth in material description, logical presentation of report contents, and effective report planning. The "report preparation" category calls for such items as:

preparing an outline; writing a draft; and, making revisions to the draft copy, as required. The "data presentation" category includes: only using the necessary data in the main body of the report, and placing the rest in an appendix; paying careful attention to the form in which data are presented effectively in the text (for example, sketches, block diagrams, charts, drawings); and remembering that diagrams, graphs, and charts generally make a stronger visual impression than tables. Some of the "avoidance items" are: poor reproduction; incorrect references; incorrect spelling; undefined symbols or abbreviations; incomplete or incorrect diagrams or table identifications and captions; disordered or incorrectly numbered pages; and ambiguous, unfamiliar, or obscure words.

1.6 FINAL REPORTS, PROPOSALS, AND PROGRESS REPORTS

If report formats are not standardized, the contents may vary significantly from one writer to another and from one project to another. It is therefore wise to establish at least minimal format standards. This section provides general formats for proposals, progress reports, and final reports in the technical arena. Note that numbering the sections of a report provides a useful guideline to both writers and readers.

1.6.1 FINAL REPORTS

The basic elements of a final technical report are as follows:

1. **Front matter.** The front matter precedes the text of the report and includes the following:
   - Title page.
   - Author(s)–affiliations, necessary signatures, date of publication. (The title and other information are usually on the first page of a report.)
   - Summary–Summarizes the contents of the report and is generally on a separate page.
   - Preface–Provides brief background information and is on a separate page.
   - Acknowledgments (if applicable)–May be part of the preface or by itself.
   - Table of Contents
   - List of Figures, including their corresponding page numbers.
   - List of Tables, including their corresponding page numbers.

2. **Introduction.** This section provides the detailed background information required to understand the subject of the report. A summary of the literature review may form part of this section. In some cases, the "introduction" may become Chapter 1 of the report.

3. **Discussion.** This section of the report contains the comprehensive explanation of the subject under study, from the initial proposal stage through the detailed analy-

sis of the findings to the presentation of the results. The discussion may be divided into several chapters, such as: Chapter 2, Equipment Assembly or Setup; Chapter 3, Approach; Chapter 4, Results; and Chapter 5, Analysis of Results. In turn, each chapter may also be divided into several sections, such as 2.1, 2.2, 2.3, and so on.

4. **Conclusions.** Covered in a separate chapter, the conclusions present the inferences and direction indicated by the results analysis.

5. **References.** The reference or bibliography section may be placed after the appendix section. This section lists all the documents to which the writers referred. The information on each document must be complete and must follow the same format throughout. Although the format may differ from one company to another, the format should be consistent within a single report. Examples of listings for books, journal articles, articles published in conference proceedings, and technical reports are as follows:

   a. **Books.**
      1. Smith, A. K., *Technical Report Writing,* John Wiley & Sons, Inc., New York, 1992.
   b. **Journal articles.**
      2. Jones, S.D., "A Study of Modern Engineering Design Processes," *Journal of Mechanical Engineering,* Vol. 50, 1990, pp. 15–25.
   c. **Articles published in conference proceedings.**
      3. Williams, R., Engineering Design Problems, *Proceedings of the Annual American Society of Mechanical Engineers Conference,* 1991, pp.105–114.
   d. **Technical Reports.**
      4. Proctor, C.L., A Reliability Study of Mechanical Pumps, Report No. SR120, 1990. Available from the Department of Mechanical Engineering, University of Ottawa, Ottawa, Ontario, K1N 6N5, Canada.

6. **Appendices.** This includes material deemed to be beyond the scope of the main body of the report. Again, the appendix section may be divided into as many subsections as necessary. For example, we could have: Appendix I, Detailed Analyses; Appendix II, Computer Programs; Appendix III, Numerical Tables; and Appendix IV, Figures.

1.6.2 PROPOSALS

Technical proposals are generally written in a predefined format, and are often done in response to a document called a Request for Proposals. Although the specific elements of a proposal may vary from one application to another, the most frequent ones are as follows:

1. **Front matter.**
   a. Title.
   b. Author(s) and affiliation, necessary signatures (if applicable), date of proposal.
   c. Company, department, or agency to whom the proposal is written.

2. **Summary.** This includes synopsis of the problem or situation, in addition to a brief abstract of the proposed solution(s) and recommendation(s).

3. **Introduction.** This section describes the situation or problem to be solved, establishes why it is a problem, and clearly states what must be improved or resolved.

4. **Proposal details.** The proposal details are divided into three general areas:
   a. **Proposed solution.** This section contains a detailed explanation of the proposed method for solving the problem, and includes the following:
      - Benefits and drawbacks of the proposed solution.
      - Cost of the proposed solution.
      - Time breakdown associated with the proposed solution.
      - Available resources.
      - Competence/qualifications of those proposed to perform the work.
      - Results or improvements to be achieved by the proposed solution.
      - Explanation as to how the proposed solution is to be implemented.
   b. **Alternative approaches.** Occasionally, more than one approach may be applicable to the solution of the problem. Alternative approaches must clearly state any differences in underlying assumptions, or conditions under which these approaches would apply.
   c. **Evaluation.** This section is concerned with the analysis of all proposed solutions, including a comparison. The most suitable approach should be clearly identified.

5. **Recommended action.** This section presents a clear, positive, and strong statement or a direct request.

6. **Supporting data.** This section of the proposal includes test results, detailed analyses, and other related material to support the earlier sections of the proposal.

1.6.3 PROGRESS REPORTS

The purpose of these documents is to inform others regarding the progress of a specific task or project. Usually, a progress report is made up of several elements, which can vary from one report to another. The following components normally comprise a progress report:

1. **Front matter.**
   a. Title.
   b. Author(s) and affiliation(s), necessary signatures (if applicable), date of the report.
   c. Dates of work period covered by the report.

2. **Summary.** This section briefly describes the overall situation.

3. **Background.** This section covers the situation and events leading up to the report.

4. **Progress.** This section concentrates on such items as: work accomplished, problems encountered, and the effects of any problems on the progress.

5. **Present situation.** This section describes the work activities currently being performed.

6. **Future plans.** This section discusses plans for completing the work, and indicating when it is projected to be accomplished.

7. **Supporting data.** The information in this section supports the previous statements made in the report.

1.7   ILLUSTRATION GUIDELINES

Visual elements, such as figures, charts, graphs, etc., in technical reports have specific purposes and convey specific information. Visuals are used to explain, illustrate, demonstrate, verify, or support the written material. Visuals or illustrations are used in the following circumstances [5]:

1. The process can be more clearly explained visually.
2. The reader's understanding may be limited.
3. Speed is absolutely necessary, and reading is too slow.
4. There is a need to generate readers' interest in the subject matter.

However visuals are only valuable if their presentation is effective. The general guidelines for preparing effective visuals (illustrations) are as follows [5,9]:

1. Reference all illustrations in the text.
2. Reference the data source, as well as the graphic designer.
3. Carefully plan the placement of illustrations. Preferred positions are: the page opposite the text reference; the page following the first text reference; the same page as the text reference; or an appendix.
4. Specify all units of measure and the scale used in the drawing.
5. Label each illustration with an identifying caption and title. Include the figure source, if it was obtained from another document.
6. Try to place illustrations in the vertical position.
7. Spell words out rather than using abbreviations.
8. When a document has five or more figures, include a List of Figures in the front matter.
9. Surround the visual with sufficient white space to separate it from the written text.
10. Avoid putting too much on an illustration.

Useful guidelines for several types of illustrations are presented here.

1.7.1 DRAWINGS

Drawings are generally used to display the actual appearance of an object. Some drawings are very simple, while others are quite complicated. In technical work, the drawings may be cutaway views, perspective drawings, action views, exploded views, phantom views, etc. Some of the guidelines for technical drawings are: select the right amount of detail, select the most relevant view, label all parts, and use legends if there are a significant number of parts.

1.7.2 TABLES

Tables are useful for displaying various types of data and may be classified as either formal or informal. A table should be presented on a single page whenever feasible; each column and row should be clearly labeled; numbers should be limited to two decimal places unless greater accuracy is required; columns to be compared should be placed next to each other; the numbers in a column should be aligned by the decimals; standard symbols and units of measure should be used; and footnotes should be used to clarify or explain headings and entries in the table, if necessary.

1.7.3 CHARTS

Charts are used to represent steps, components, or the chronology of an organization, mechanism, object, or organism. Examples of charts are: organizational charts, flowcharts, bar charts, pie charts, schedule charts, and line charts. Guidelines for each of these chart types are presented here.

Organizational Charts   These charts are used to display the structure of an organization in terms of its people, their positions, or the work units. Some suggestions for developing organizational charts are: use boxes to represent positions; connect the boxes with solid lines, for direct reporting relationships, or dotted lines for indirect or staff relationships; use various shapes (if possible) to indicate different levels or types of jobs; and position the boxes to minimize or eliminate line crossovers.

Flowcharts   These are used to show the sequence of steps in a process, and they frequently indicate the time required for each step. Some of the guidelines associated with flowcharts are: limit and clearly label the number of different shapes; present only overviews; use sufficient space for the chart to be easy to follow; use a legend when necessary; and run the sequence of events either from left to right or from top to bottom.

Bar Charts These are used to present information for comparison, and the bars may run either horizontally or vertically. Some of the rules for developing bar charts are: arrange the order of the bars for greatest effectiveness; limit the number of bars; use different shadings for different bars; and keep the bar widths equal.

Pie Charts   These charts are used to depict approximate percentage relationships between individual components or elements of a complete system or concept, such as a budget. Some of the guidelines for developing effective pie charts are: use as few divisions as possible; if necessary, combine two or more elements into one wedge of the pie; draw and label the wedges carefully and clearly; and move clockwise from 12:00 and from the largest to the smallest wedge.

Schedule Charts   These charts are used to show when specific activities are to be completed, especially in proposals and feasibility and progress reports. These charts are also referred to as milestone or Gantt charts. Some of the rules for developing schedule charts are: include only the pertinent activities; if necessary for clarity, use one main chart for the principal activities and several ancillary charts for the supporting activities; create a format to fit the items being depicted; list the activities in sequence, beginning at the top of the chart; be as realistic as possible regarding the schedule; label each step clearly and concisely; and run the labels in the same direction as the chart.

Line Charts   These charts are used to show trends in the changes of two related variables. Examples of line chart applications are stock price trends, car prices, and weather changes over time. The guidelines for developing effective line charts are: avoid placing numbers on the chart itself unless absolutely necessary; use chart lines that are a different thickness than the axis lines; place the charts for maximum effectiveness with respect to the corresponding text; and avoid having too many lines on one chart.

1.7.4 PHOTOGRAPHS

Photographs are used to show the actual appearance of an object, mechanism, or organism. They are particularly useful when the emphasis is on realism and when the subject is to be shown in its natural setting. When photographs show unnecessary detail, callouts can be used to draw attention to the desired features. Further emphasis of a specific element can be accomplished with reduction, enlargement, or cropping of the photographs.

1.7.5 DIAGRAMS

Diagrams are used to show the physical elements of objects, mechanisms, or organisms. In diagrams, complex structures are represented without unnecessary details, which means that diagrams are usually easier to understand than more complex drawings.

1.8 READABILITY INDICES

Today, the term "readability" is used as a measurable aspect of writing. Several people such as Flesch-Kincaid, Fog, and Fry [5] have developed readability formulas or indices. The main objective of these formulas is to provide feedback to writers on whether their work is acceptably easy to understand, or difficult to follow. The formulas are based on the relationships between the mean sentence length (that is, number of words per sentence) and mean word length (that is, number of syllables per word). The reasoning that forms the basis for these formulas is: the greater the mean number of syllables per word and the higher the mean number of words per sentence, the more difficult the written document is to understand. These formulas estimate a document's readability through two avenues: a grade level (grade one to graduate study), and a level of difficulty (easy to difficult).

Some of the difficulties in using readability formulas are:

1. The difficulty of the content is not considered.

2. The use of subordination to increase comprehension is ignored.

3. The results obtained from different readability formulas cannot be compared because of inconsistencies in the formulas.

1.8.1 FLESCH-KINCAID INDEX

The final figure obtained from this formula is the education grade level required to understand the written material. The index is defined as follows [5]:

$$GL = (0.39)AW + (11.8)AS - 15.59 \qquad \textbf{[1.1]}$$

where

GL   is the grade level.

AW   is the average number of words per sentence.

AS   is the average number of syllables per word.

Example 1.1

Assume that a technical report contains an average of 13 words per sentence and an average of two syllables per word. Compute the readability grade level of the report, using the Flesch-Kincaid index.

Substituting the given data into Equation (1.1) yields

$$GL = (0.39)(13) + (11.8)(2) - 15.59$$

$$\simeq 13$$

Therefore, the years of education required to understand the report is 13.

1.8.2 FOG FORMULA

Again, this index produces a figure representing the number of years of education required to comprehend the material. The Fog index is expressed by

$$FI = (0.4)[AW + WS] \qquad \textbf{[1.2]}$$

where

FI   is the Fog index value.

AW   is the average number of words per sentence (always take a
       sample of at least 100-word length).

WS   is the number of words of three syllables or more per 100 words. In computing the value of WS, avoid counting: capitalized words, combinations of short, easy words, such as "park keeper," and verbs that are three syllables because "es" or "ed" are added.

1.9 PROBLEMS

1. Discuss the major steps involved in writing a technical report.

2. Discuss the types of editing that a technical document may require.

3. Identify and describe the information sources that can be used in writing a technical report.

4. Describe at least six types of technical documents.

5. What are the qualities of a good report?

6. Describe each element of a final technical report.

7. Briefly discuss the process of listing references at the end of a document. Give examples for listing books, journal articles, conference proceeding papers, and technical reports.

8. Discuss the guidelines for preparing the following:
   • Flowcharts.
   • Drawings.
   • Tables.
   • Pie charts.

REFERENCES

1. Olsen, L. A.; and T. N. Huckin. *Technical Writing and Professional Communication.* New York: McGraw-Hill, 1991.

2. Davis, R. M. "Technical Writing: Who Needs It?" *Engineering Education,* 1977, pp.209–211.

3. Gilbert-Storms, C. "What Business School Graduates Say About the Writing They Do at Work: Implications for the Business Communication Course." *The ABCA Bulletin,* December 1983, pp. 13–18.

4. Spretnak, C. M. "A Survey of the Frequency and Importance of Technical Communication in an Engineering Career." *The Technical Writing Teacher,* 1982, pp. 133–136.

5. Burnett, R. E. *Technical Communication.* Belmont, CA: Wadsworth Publishing Co., 1990.

6. Pliny. *Natural History* X,Translated by D. E. Eichholz. Cambridge, MA: Harvard University Press, 1962.

7. Eisenberg, A. *Writing Well for the Technical Professions.* New York: Harper & Row, 1989.

8. Blicq, R. S. *Technically-Write!* Scarborough, Ont.: Prentice-Hall Canada, Inc., 1983.

9. Pfeiffer, W. S. *Technical Writing: A Practical Approach.* New York: Macmillan, 1981.

10. Houp, K. W.; and T. E. Pearsall. *Reporting Technical Information.* New York: Macmillan, 1988.

chapter

2

INTRODUCTION TO DESIGN

2.1 INTRODUCTION

In the discipline of engineering, the term "design" may convey different meanings to different people [1]. To some, a designer is a person who uses drafting tools to draw the details of a part. To others, a design is the creation of a sophisticated system, such as a computer system. For our purposes, the term "engineering design" means the design of items of a technical nature—structures, devices, etc.

Humans have been designing engineering-related objects and structures for thousands of years. The Egyptian pyramids and the great wall of China are two prime examples of ancient civil engineering structures. A treatise on architecture by Vitruvius (a Roman architect) was written in 30 B.C. Since the 17th century, many new technical objects have been designed and developed, including the following:

1. **Telescope (reflecting).** This device was designed and constructed by Sir Isaac Newton in 1669 in Great Britain.

2. **Clock (pendulum).** This was designed and constructed by Christian Huygens in 1656 in the Netherlands.

3. **Steam engine.** This was designed and developed by James Watt in 1769 in Great Britain.

4. **Submarine.** This was designed and developed by David Bushnell in 1776 in the United States.

5. **Electric generator (DC).** This was designed and constructed by Michael Faraday in 1831 in Great Britain.

6. **Motor car.** This was designed and developed by Karl Benz in 1885 in Germany.

7. **Aeroplane.** This was designed and developed by Wilbur and Orville Wright in 1903 in the United States.

8. **Helicopter.** This unique craft was designed and developed by Louis G. Breguet in 1909 in France.

9. **Sewing machine.** This was designed and constructed by Walter Hunt in 1832 in the United States.

10. **Typewriter.** This basic office machine was designed and constructed by Christopher Sholes in 1868 in the United States.

The first book on engineering drawings, entitled *Geometrical Drawings*, by William Minifie,

16

was published in the United States in 1849. Since that time, a great many textbooks and articles on design and engineering drawings have appeared.

2.2 DESIGN PURPOSES AND DESIGN FAILURES

Each year, hundreds of engineering products are designed and put into use. The reasons for their design [2,3] could be: meeting the competition, responding to social changes, developing a new approach, reducing inconvenience or hazard, or reducing cost. Occasionally, when newly built facilities or objects are used in the field, catastrophic failures occur, even though the items were apparently designed with the utmost care. In recent history, three well-publicized disasters associated with engineering systems were: the Three Mile Island nuclear power plant, the space shuttle Challenger, and the Chernobyl nuclear power plant. Previous engineering design failures were as follows [4]:

1. An offshore oil rig named the Alexander L. Kielland broke up in normal North Sea weather in 1980.
2. A Boeing 737–200 lost its cabin roof during flight in 1988.
3. A bridge over the Tacoma Narrows collapsed during a moderate 40-mph crosswind in 1940.
4. A DC-10 airplane lost an engine in flight and crashed in 1979.
5. A de Havilland Comet airplane exploded in the air and crashed in the 1950's.
6. The Hartford Civic Center roof collapsed under the weight of snow in 1978.
7. A skywalk of the Kansas City Hyatt Regency Hotel collapsed just after the hotel was opened in 1981.
8. One leaf of the bridge over the Hackensack River in New Jersey collapsed within two years of being placed in 1928.

After investigation, failures such as those mentioned have resulted in the identification of possible reasons or sources. Engineering designers must be aware of such sources for design failures in order to avoid or at least minimize future problems with newly designed items. Common sources of engineering failures, errors, and omissions are as follows [4]:

- Incorrect or overextended assumptions.
- Poor understanding of the problem to be solved.
- Incorrect design specifications.
- Faulty manufacture and assembly.
- Error in design calculations.
- Inadequate data collection and incorrect or incomplete experimentation.
- Errors in drawings.

- Errors in packaging and shipping.
- Incorrect storage and/or poor storage facility.
- Incorrect installation and use.
- Faulty reasoning from good assumptions.

2.3 DESIGN TYPES

There are different types of design that a design engineer may be expected to produce. These demand varying degrees of ability and creativity. In general, the designs may be classified into three main categories [5]:

1. **Creative design.** This is the design of a totally new product without any precedent whatsoever. This type of design work requires a high degree of competence. Relatively few design engineers will be employed in this type of design activity.

2. **Adaptive design.** This is the adaptation of existing designs to meet new purposes. A major portion of the design work undertaken by engineers falls in this category. This type of activity generally requires at least basic technical skills and some level of creativity.

3. **Developmental design.** To a certain degree, this is also the adaptation of an existing design, but only as a basis. This type of design work may also involve a considerable amount of technical work, and the final product could be quite different from the original one.

2.4 ENGINEERING DESIGN MANPOWER

Depending on the type and size of the design effort, several professionals could be directly or indirectly involved in completing the design task. Also, their specialties may vary from one design project to another. In addition to the design engineer, other individuals who might be associated with design work are as follows:

1. **Reliability engineer.** This person evaluates the design for projected reliability of the end product. Sometimes, a reliability parameter, such as the *mean time between failure (MTBF),* may be given in the product design specification. The reliability engineer ensures that the specified value of the parameter is fully satisfied by the end product.

2. **Maintenance/maintainability engineer.** This professional evaluates the design for maintenance and maintainability. Again, in some cases, in the product's design specification, the *mean time to repair (MTTR)* may be defined, and this engineer ensures its fulfillment by the final product.

3. **Quality control engineer.** This individual evaluates the design from the quality control aspect especially when the designed item goes into production.

4. **Human factors engineer.** This person ensures that adequate consideration is given to the human element during the design. Human factors analysis during the design phase contribute to the effectiveness of the product's operation and maintenance.

5. **Safety engineer.** This professional evaluates the design from the standpoint of safety by performing various safety-related analyses.

6. **Manufacturing engineer.** This specialist evaluates the design to determine its ease of manufacturability.

Additional specialists, such as a financial analyst, tooling engineer, drafting supervisor, value engineer, component engineer, and customer engineer, may also participate in the design process, depending on the nature of the project.

2.4.1 QUALITIES OF A TYPICAL DESIGN ENGINEER

A design department generally interfaces with many other departments or groups (such as research, development and testing; production; marketing; finance; operation and maintenance; commissioning; legal; safety; technical support services; and customers). Some of the tasks performed by the design engineer are: designing the product, participating in design reviews, optimizing the design, keeping up to date with the changing environments and technology, coordinating the design with other concerned people, generating new ideas for designs, keeping the design within specified constraints, answering questions regarding the design, keeping management informed, and keeping records of all changes. To interface effectively with such diverse groups and perform these tasks competently, the design engineer must possess many qualities. Some of these are as follows [2,5,6]:

1. **Excellent scientific knowledge.** A good designer must have an excellent scientific knowledge in the area of specialization in order to produce satisfactory designs. Basic subjects in which the designer should be well versed are: physics, chemistry, technical drawing, elementary and advanced mathematics, engineering and technology, economics, and management.

2. **Ability.** In design work, it is important to be able to think logically, both during the actual design process and in the preparation of the overall design strategy.

3. **Creative and innovative mind.** During designing, a designer generally faces various types of problems: technical, managerial, economic, etc. Addressing these problems requires a creative and innovative mind.

4. **Conceptual ability.** A good designer must have the capability to comprehend and visualize the concept of the design.

5. **Good communication capability.** A good designer must be able to communicate effectively both verbally and in writing, so that the desired information can be disseminated to other interested parties.

6. **Good personal qualities.** These include good memory, concentration capability, perseverance, integrity, willpower, and temperament, ability to tolerate criticisms and flexibility.

2.4.2   DESIGN REVIEW BOARD CHAIRMAN

The Design Review Board Chairman plays a crucial role in the success of the design, especially during the design review phase. Functions performed by the board chairman are: chairing the design reviews, establishing the types of design reviews to be conducted, scheduling the reviews, directing the necessary follow-up action(s), preparing the agenda and the minutes and distributing these to the appropriate people, and establishing the procedure for choosing specific items for review. To accomplish these tasks, the board chairman should: be skilled in leading a technical team; be tactful and have a broad understanding of the technical problem; have a high degree of discretion, good personality, and be totally objective with respect to the proposed design [6,7].

2.5   DESIGN-RELATED INFORMATION SOURCES

A design engineer may require various types of information, which can be obtained from many sources. It is reported that the world of technical literature is doubling every 10 to 15 years [8], which translates to roughly 2 million technical articles per year. The sources of information for engineering design may be divided into two broad categories: public or private. Typical public sources include:

1. Federal Government departments and agencies, including such organizations as defense, energy, commerce, and science and technology.

2. Local and state government departments; including transportation, consumer affairs, and building code regulations.

3. Research organizations, universities, colleges, and museums.

4. Libraries, including university, community, and other facilities.

5. Foreign embassies and high commissions.

Typical private information sources include:

1. Nonprofit bodies and services, which are divided into three main categories: professional societies, membership organizations, and trade and labor associations.

2. Profit-oriented organizations, including manufacturing and user organizations, and consultants. Types of information available from these organizations include: catalogs, test and cost databases, and operation and maintenance databases.

3. Individuals, such as experts in their fields, colleagues and professional associates, university professors, researchers, etc.

2.5.1 GOVERNMENT INFORMATION SOURCES

In the United States, there are several government or government-sponsored organizations that can provide engineering design information. Some of these are listed here.

1. Mechanical Properties Data Center
   Battelle Columbus Laboratories
   Columbus, OH 43201

2. Government Industry Data Exchange Program (GIDEP)
   GIDEP Operations Center
   U.S. Department of the Navy
   Naval Weapons Station, Seal Beach
   Corona, CA 91720

3. National Technical Information Service (NTIS)
   U.S. Department of Commerce
   5285 Port Royal Road
   Springfield, VA 22151

4. Reliability Analysis Center
   Rome Air Development Center (RADC)
   Griffiss Air Force Base
   New York, NY 13441–5700

5. Defense Technical Information Center
   Defense Logistics Agency
   Cameron Station
   Alexandria, VA 22314

6. NASA Space Science Data Center
   Goddard Space Flight Center
   Greenbelt, MD 20771

7. Machinability Data Center
   Metcut Research
   Cincinnati, OH

8. NASA Parts Reliability Information Center (PRINCE)
   George C. Marshall Space Flight Center
   Huntsville, AL 35812

9. Data on Trucks and Vans
   Commanding General
   Attn: DRSTA-QRA, U.S. Army
   Automotive-Tank Command
   Warren, MI 48090

2.5.2 LIBRARY INFORMATION SOURCES

A library is a good source of information for open or unclassified published literature. The vareity and type of literature available depends on the nature of the library. Design information sources in libraries may be classified into such categories as [8]:

1. Handbooks.
2. Encyclopedias and technical dictionaries.
3. Indexing and abstracting services.
4. Standards.
5. Catalogs and manufacturers' brochures.
6. Technical reports and bibliographies.
7. Textbooks.
8. Professional journals and newsletters.

Handbooks, available on various technical areas, play a vital role. Some of these are as follows:

1. *Mechanical Engineers' Handbook*. ed. M. Kutz. New York: Wiley, 1986.
2. *Electronics Engineers' Handbook*. ed. D.G. Fink, and D. Christiansen. New York: McGraw-Hill, 1982.
3. *Handbook of Engineering Fundamentals*. ed. M. Souders, and O.W. Eshback. New York: Wiley, 1975.
4. *Maintenance Engineering Handbook*. ed. L.R. Higgins, and L.C. Morrow. New York: McGraw-Hill, 1977.
5. *Pump Handbook*. ed. I.J. Karassik. New York: McGraw-Hill, 1976.
6. *Automotive Electronics Reliability Handbook*. AE-9, Warrendale, PA: Society of Automotive Engineers, 1987.
7. Parmley, R.O. *Standard Handbook of Fastening and Joining*. New York: McGraw-Hill, 1977.
8. *Handbook of Engineering Management*. ed. J.E. Ullman. New York: Wiley, 1986.
9. *Structural Engineering Handbook*. ed. E.H. Gaylord, and G.N. Gaylord. New York: McGraw-Hill, 1979.
10. *Mathematical Handbook for Scientists and Engineers*. ed. G.A. Korn, and T.M. Korn. New York: McGraw-Hill, 1968.
11. *Standard Handbook for Electrical Engineers*. ed. D.G. Fink, and H.W. Beaty. New York: McGraw-Hill, 1978.
12. *Piping Handbook*. ed. S. Crocker, and R.C. King. New York: McGraw-Hill, 1967.
13. *Standard Handbook for Civil Engineers*. ed. F.S. Merritt. New York: McGraw-Hill, 1983.

14. *Agricultural Engineers' Handbook.* ed. C.B. Richey. New York: McGraw-Hill, 1971.

15. *Tool and Manufacturing Engineers' Handbook.* ed. D.B. Dallas. New York: McGraw-Hill, 1976.

Encyclopedias and technical dictionaries can also provide useful information for designers. Some examples of encyclopedias and technical dictionaries are:

1. *Encyclopedia of Computer Science and Technology.* ed. J. Beizer. New York: Marcel Dekker, Inc., 1980.

2. *Encyclopedia of Engineering Materials and Processes.* ed. H.R. Clauser. New York: Van Nostrand Reinhold, 1963.

3. *McGraw-Hill Dictionary of Scientific and Technical Terms.* ed. S. Parker. New York: McGraw-Hill, 1984.

4. Horner, J.G. *Dictionary of Mechanical Engineering Terms.* New York: Heinman, 1967.

5. *Encyclopedic Dictionary of Mathematics for Engineers and Applied Scientists.* ed. I.N. Snedden. London: Pergamon Press, 1976.

6. *McGraw-Hill Encyclopedia of Science and Technology.* New York: McGraw-Hill, 1982.

7. *McGraw-Hill Dictionary of Science and Engineering.* New York: McGraw-Hill, 1984.

The past few decades have seen an explosive growth in scientific literature worldwide. For example, in 1960, there were 18,800 scientific journals; in 1980 that figure had grown to 62,000. Indexing and abstracting services provide information on the latest periodical literature. Indexing services cite articles by title, author, and bibliographic data [8]. Abstracting services also provide a synopsis of the article. Examples of indexing and abstracting services are as follows (the computer access name is listed in brackets, if available):

1. Engineering Index (COMPENDEX).
2. Science Abstracts (INSPEC).
3. Sciences Citation Index (SCISEARCH).
4. International Aerospace Abstracts.
5. Applied Mechanics Reviews.
6. Index to Scientific and Technical Proceedings.
7. Energy Index (ENERGYLINE).
8. Applied Science and Technology Index.
9. Building Science Abstracts.
10. British Technology Index.

Many libraries provide access to various types of standards applicable to designing. A standard establishes technical limitations and applications for materials, items, methods, designs, processes, and engineering practices [8]. The basic objectives of a standard are to meet the needs of designers and to control variations. At present, there are over 60,000 government and private industry standards and specifications in use in the United States alone. The developers of such standards and specifications include: the United States Department of Defense (DOD), American National Standards Institute (ANSI), professional and trade associations and societies, and the International Organization for Standardization (ISO). The names and addresses of some of these organizations are as follows:

1. American Society for Testing and Materials
   1916 Race Street
   Philadelphia, PA 19807

2. American National Standards Institute
   10 East 40th Street
   New York, NY 10016

3. U.S. Government Printing Office
   North Capitol and H Streets, NW
   Washington, D.C. 20402

2.5.3 THE INFORMATION SUPERHIGHWAY

Advances in electronic communications and related technologies are leading the way to the development of the *information superhighway*, which may have many different meanings, depending on the context. There is currently no single definition of what the information superhighway will be or what will travel on it [9], even though the superhighway scheme was proposed by Vice President Al Gore as early as 1979.

In addition to the communications infrastructure, the information superhighway has three key components [10]:

1. **Information providers.** This includes digital libraries, local broadcasters, individuals, information services, and so on.
2. **Software**. This allows easy and effective utilization of the services and information available.
3. **Information appliances**. These include computers, televisions, telephones, and so on.

The information superhighway has a wide range of applications, for both public organizations and the private sector. For public institutions, the superhighway will improve the delivery of services and the dissemination of government information. In the private sector, the highway will facilitate the formation of more effective teams with members at different locations, as well as providing easy access to information on just about any subject imaginable. As an example, on-line databases, at-home shopping, and electronic books are already available to some degree.

The information superhighway is already in place in the form of the *Internet*, the world's largest computer bulletin board and data bank. Millions of people around the world use the Internet to share research results, send and receive mail, play games, etc. Recently, the United States Government committed $400 million for the development of the *National Research and Education Network (NREN)*, which will initially use the existing U.S. Internet. During the early phase, NREN will connect over a dozen leading research centers in the United States through a gigabit-per-second [11] fiber-optic-based network, using existing fiber-optic cables. Ultimately, NREN will displace the current Internet, which is rather slow, for transmitting large texts such as complete books, high-resolution graphics, and extensive databases. In addition, the information superhighway is expected to play a leading role in the field of engineering design.

Internet   This is the oldest long-distance computer network in the United States and the largest of its kind in the world. The Internet provides the mechanism for computer centers located at various sites to communicate and to share services and sources effectively. The term "internet" may mean different things to different people [9]; for example, the world's largest computer network, a massive global information service, a set of standards for data communication, a marketplace without boundaries, or the world's largest pen pal system.

Soon after the development of the computer, the need to transfer information from one machine to another became essential. In the early days, this was accomplished by "writing" the information to an intermediate medium, such as punched cards or magnetic tape, and then physically transporting that medium to the other machine. To improve this task, computer scientists in the early 1960s started to explore ways of directly connecting remote computers.

In the mid-1960s, the United States Government funded an experimental project for this purpose, called the *ARPANET* [12]. This project led to the development of the software protocol called *Transmission Control Protocol/Internet (network) Protocol (TCP/IP)*. The computers connected to the network used TCP/IP to communicate with each other. In the 1970s, TCP/IP became the ARPANET standard network protocol, and the early 1980s witnessed the conversion of all interconnected research networks to TCP/IP. ARPANET thus became the backbone of the new internet, which is comprised of all TCP/IP-based networks connected to ARPANET by the end of 1983. The widespread use of the Internet is witnessed by the increase in the number of hosts (computers that provide access to Internet services) from 235 in May of 1982 to approximately 3.2 million in July of 1994.

The Internet is not the property of any organization, in the usual sense of the word, but the National Science Foundation (NSF) provides the basic funding in the United States. Also, the *Internet Engineering Task Force (IETF),* composed of experts on Internet issues, provides technical support. Various organizations involved with Internet-related education and other issues include: Internet Society (ISOC), Computer Professionals for Social Responsibility (CPSR), Corporation of National Research Initiatives (CNRI), etc. Information on these organizations is given in Reference 4.

People use the Internet for a variety of purposes [9]:

1. Send and receive messages globally through electronic mail (E-mail).
2. Put out newsletters.
3. Share common-interest information with their peers.
4. Access files and data.
5. Establish Internet resources.

A direct link to the Internet may or may not be useful to everyone. Factors that should be considered before establishing such a link are as follows [9]:

1. Technology perspective versus strategic perspective.
2. Internet opportunities for the organization.
3. Competitors presence on the Internet.
4. Realistic expectations.
5. Internet limitations.
6. Limitations of in-house infrastructure.

In its current state, the Internet may not provide all the information required by an individual or organization. It can even be frustrating to search through the information that the Internet does provide. Some of the difficulties associated with the Internet are as follows:

1. Since the Internet had its roots in research and academia, the extensive information available tends to be more suitable for those areas.
2. The accessible information sources on the Internet may not be permanent.
3. There is no "master information index."
4. The reliability of information on the Internet is sometimes questionable.

In addition to these problems, many of the originally available services lack effective documentation and are rather difficult to use. Now however, the Internet has been opened up to commercial and private users, and more "user friendly" services are being developed, in addition to new easy-to-use interfaces to the older services. Some of the well-known Internet services, or information retrieval tools, are discussed in the following paragraphs [9, 12].

Gopher   The Gopher protocol and software package allows the user to browse information systems. Generally speaking, Gopher is simple to use. The name Gopher is derived from the school mascot of the University of Minnesota, which developed the package.

Gopher may be described as an express elevator in the Internet that can efficiently cut through layers of information, as well as allowing the user to traverse the world's data banks. Gopher can also connect various major Internet computers together into a single, unified information service.

Gopher can help the user locate many Internet resources, including:

1. Information relating to Internet resources.
2. Comprehensive guides to the Internet, as well as documents on using various Internet services.
3. Software locations within specific *file transfer protocol (FTP)* sites.
4. Specific mailing lists.

Wide Area Information Server (WAIS)   WAIS is useful for finding documents available on the Internet on a specific subject. The WAIS searches indexed databases with keywords and then lists the addresses of locatable documents on the subject of interest. An important feature of WAIS is that it allows the use of the client's software running on the local computer to ask for information in a straightforward language similar to English. The WAIS system has many features; one of the important ones is that a WAIS server possesses indexes pointing to other WAIS servers. The indexes to all known Internet WAIS servers are maintained by a central site on the Internet.

World Wide Web (WWW, W3, or Web)   The Web is one of the latest client–server based Internet services and was developed by the European Laboratory for Particle Physics (CERN) in the late 1980s. The Web is an Internet initiative that promises relatively easy browsing of Internet resources as well as a type of indexing of the information available on the Internet. In addition, the Web allows the combination of text, graphics, audio, and animation to create a document that can provide a total learning experience. Leads within Web documents can efficiently take the user to other relevant documents.

In essence, the Web is the closest the Internet has come to a comprehensive, user-friendly interface. Recently, a number of books on the Web have been published. *Using the World Wide Web* [13], published by Que is a particularly useful document for better understanding and effectively using the Web. References 9 and 12 are also useful in this regard.

Mosaic   This client software package was developed by the National Center for Super Computing Applications (NCSA). During the early part of 1993, the package was released to the Internet. Mosaic is an easy-to-use interface for traversing the

Web. It possesses features that help the user keep track of locations visited and also helps in returning to them relatively quickly. The combination of Mosaic and WWW is called *hypermedia*, which is the combination of hypertext and multimedia.

2.6 THE ROLE OF SOCIETAL CONSIDERATIONS IN DESIGN

A "society" is an organization of people having some common traits and living within specific boundaries. A successful engineering design must carefully evaluate the society's needs. The operation of various engineering systems may directly or indirectly affect daily life. Examples are: electric power generation and distribution systems, transportation systems, and telecommunications systems. In addition, various social forces can impact the engineering profession: consumer rights, the antinuclear movement, occupational health and safety, environmental protection, etc.

Some of the ways in which increased societal awareness of technology has influenced the practice of engineering are as follows [14]:

1. Increase in the time required to plan and study the potential impacts of engineering projects.
2. Increase in the input needed from the legal profession concerning engineering decisions.
3. Increase in the cost of "defensive studies" to protect companies against possible litigation.
4. Increase in the amount of spending on "overhead" areas, such as environmental control and safety.
5. Increase in government regulations and monitoring.

The growing importance of technology to the society at large has led to the development of a methodology called *technology assessment (TA)*. This methodology systematically determines the impact of technology on the physical, social, political, and economic environment. To implement the TA methodology, the United States Congress established the Office of Technology Assessment. TA practitioners try to provide the following:

1. The decision-making tools for allocating resources and developing technological priorities.
2. An early warning system for environmental incidents.
3. The necessary monitoring and surveillance mechanisms.

The characteristics that differentiate TA from traditional engineering planning methods are: the use of an interdisciplinary methodology; a close association with policymaking, rather than technical problem solving; a consideration of the needs of a wide range of constituencies; and a concern for investigating second, third, and higher-level impacts that may not be examined in routine engineering analysis.

2.6.1 RISK

Risk is a critical element of societal considerations in engineering design. Risk is the potential for creating an undesirable consequence of an event. Six general categories of societal hazards are as follows [15]:

1. **Large, complex technological system failures.** These include the failure of powerplants, ships, aircraft, dams, buildings, etc.
2. **Low-level delayed-effect hazards.** These include such items as noise, microwave radiation, and asbestos poisoning.
3. **Infectious and degenerative diseases.** Examples of such diseases are influenza and heart disease.
4. **Sociopolitical disruption.** Two examples are oil embargoes and terrorism.
5. **Natural disasters.** These include floods, hurricanes, and earthquakes.

There are also a number of ethical issues relating to technological growth and risk assessment. These are: maintaining competence for effectively evaluating risks; accurately estimating the risk, using specialized technical knowledge and skill; comprehending the inputs of other risk assessment professionals; communicating with others concerning the risk assessment role in decision making; and providing input concerning the estimated risk precision and accuracy.

In 1969 in the United States, the probability of a fatality, per person per year, for accidents by motor vehicle, air travel, lightning, and fire were $3 \times 10^{-4}, 9 \times 10^{-6}, 5 \times 10^{-7}$, and 4×10^{-5}, respectively. In 1975, the cost in dollars per fatality averted for taking such societal actions as installing sulfur scrubbers in coal-fired powerplants, installing collapsible automobile steering columns, and developing and installing improved highway guardrails were $500,000, $100,000 and $34,000, respectively.

To minimize various types of technologically related risks, the United States Government has established several organizations: Environmental Protection Agency (EPA), Nuclear Regulatory Commission (NRC), Consumer Product Safety Commission (CPSC), Federal Aviation Agency (FAA), and Occupational Safety and Health Administration (OSHA).

2.6.2 QUALITY OF LIFE

Engineering design may impact the quality of life in any number of ways, for example, system analysis and synthesis may affect the quality of life positively (i.e., efficiency, resulting in reduced consumptions of material and human resources) and negatively, (i.e., pressures for centralized controls to avoid sub-optimizations, yielding less freedom and diversity). Quality of life may be defined as that characteristic which makes life desirable. It is composed of those aspects of life that are regarded and valued [16]. Such valued components may include a place to live, leisure time enjoyment, education facilities, jobs, health related facilities, and so on. In fact, L. Scheer in Reference 17 reported using 25 different factors in comparing the quality of life in 15 countries. These included: physicians per unit of population, unemployment as a percentage of the labor force, private cars per unit of population, life expectancy of a newborn male/female, per capita gross national product increase in constant prices, homicides per unit of population, fatal traffic accidents per unit of population, telephones per unit of population, television sets per unit of population, average number of persons per room, primary school children per teacher, excess consumption of calories, average work week of full-time workers in manufacturing (in hours), and percentage of women as university-level students.

C. C. Abt, in Reference 16, proposed 14 indexes to measure "obvious" quality of life factors. Some of these were as follows:

1. **Employment security.** The proposed index was [unemployment]$^{-1}$.
2. **Physical safety and security.** The proposed index was [accident and crime rate]$^{-1}$.
3. **Quiet.** The proposed index was [decibels above mean library ambient]$^{-1}$.
4. **Health.** The proposed indexes are mean mortality and [days away from work]$^{-1}$.
5. **Interesting and rewarding work.** The proposed index was job advancements over job changes.

2.7   TYPICAL UNDERGRADUATE DESIGN PROJECTS

For many years, the author has taught an undergraduate design course. In that course, groups of 5–6 students have been allowed to select their own design topic (one topic per group) as a course assignment (with some guidance from the instructor). Most of the time, the student groups successfully completed their design projects. In fact, over the years, some of the groups successfully competed in various design competitions. Examples of the design projects selected by the students are as follows:

- wheelchair traction.
- universal snooker rake.
- sheet clamp.
- hydraulic wrench.
- bicycle brake.
- solar blanket roller.
- stove switch for arthritic persons.
- bicycle lock.
- portable bicycle engine.
- water flusher.
- patient lift.
- industrial dust extractor.
- baseball pitcher.
- one-foot drive wheelchair.

- can packaging system.
- nut cracker.
- pool cleaner.
- drafting chair.
- mouth aid.
- multispeed wheelchair.
- hydraulic bicycle gear.
- brush cleaner.
- polar plotter.
- safety valve.
- exercise system.
- solar airconditioner.
- baby seat.
- chemical mixers.
- reciprocating snow machine.
- atlas rack.
- bicycle air pump.
- all-weather bicycle drivetrain.
- automatic door.
- marine aquariums.
- solar pool heater.
- compact stepladder.
- electric car window deicer/snow remover "The icebuster".
- car-wheel fastener.
- hydraulic bicycle seat.
- mechanical decoy recoil mechanism.
- rehabilitation walking aid.
- wind-powered irrigation system.
- bottle capper.
- one-arm drive wheelchair.
- crutch tip.
- automatic dog feeder.
- spraying can nozzle press.
- portable drafting kit.
- wheelchair brakes.
- snow removal device.
- bicycle cart.
- car roof rack.
- one-hand can opener.
- bicycle seat.
- super chair.
- log splitter.
- lid opener.
- doorknob handle.
- wheelchair-to-toilet transfer.
- variable-pitch propeller.
- boat pump.
- wheeled walker.
- multipurpose handyman.
- wheelchair.
- elevated seat mechanism for a wheelchair.
- wind-powered generator.
- solar blinds.
- bicycle brake system.
- tile spacer and applicator.
- lifting device.
- man-powered desert trailer.
- tree platform.
- tricycle for the handicapped.
- electric hospital cart.
- paintbrush cleaner.
- bicycle pedal basket release.
- mechanical arm.
- fire alarm for the blind and deaf.
- portable dishwasher.
- blackboard eraser.
- infrared water heater.
- goldfish feeder.

2.8 CASE STUDY: CHERNOBYL NUCLEAR POWER PLANT ACCIDENT

Seven years after the Three Mile Island nuclear accident, the Chernobyl Nuclear Power Plant accident (see Figure 2.1) in the Ukraine, formerly of the Soviet Union, occurred on April 26, 1986. At the time of the accident, the No. 4 reactor at the plant was generating only about 7 percent of its maximum output of 1,000 megawatts of power, and it was in the process of being shut down for periodic maintenance [18]. The reactor core exploded, nipping open the top of the reactor vault, and resulting in the immediate death of one worker in the refueling hall (which was on top of the reactor vault). The refueling hall was also destroyed by the explosion, and the falling debris led to the death of another worker and further damaged the reactor top.

According to a report published in the August 1986 issue of *New Scientist* [19], at a meeting organized by the International Atomic Energy Agency (IAEA) in Vienna, Professor Valery Legasov, Deputy Director of the Kurchatov Atomic Energy Institute, Moscow, conceded in his address to over 540 experts from 45 countries that there had been defects in the design of reactor types such as that used at the Chernobyl installation. In addition, he admitted that there were problems with the systems analysis and there were deficiencies in the training of the plant's operators.

2.9 PROBLEMS

1. Describe in detail the various types of designs.
2. List several common sources of engineering failures, errors, and omissions.

Figure 2.1   Chernobyl nuclear power plant installation.
| Reuters/Bettmann

3. Describe the characteristics of a good design engineer.
4. What are the typical functions of a design engineer?
5. What are the tasks of a typical design review board chairman?
6. Describe the public sources for obtaining design-related information.
7. Discuss the available design-related government information sources.
8. What are the names and addresses of at least eight organizations for obtaining engineering design-related standards?

REFERENCES

1. Shigley, J. E.; and L. D. Mitchell. *Mechanical Engineering Design.* New York: McGraw-Hill, 1983.

2. Harrisberger, L. *Engineermanship: A Philosophy of Design.* Belmont, CA: Wadsworth Publishing Co., 1966.

3. Farr, M. *Design Management.* London: Cambridge University Press, 1955.

4. Walton, J. W. *Engineering Design.* New York: West Publishing Co., 1991.

5. Ray, M. S. *Elements of Engineering Design.* Englewood Cliffs, NJ: Prentice-Hall, 1985.

6. Dhillon, B. S. *Quality Control, Reliability and Engineering Design.* New York: Marcel Dekker, Inc., 1985.

7. *Engineering Design Handbook, Development Guide for Reliability, Part II, Design for Reliability.* AMCP 706-196, Alexandria, VA: U.S. Army Material Command, (1976), pp. 11.1–11.10.

8. Dieter, G. *Engineering Design.* New York: McGraw-Hill, 1983.

9. Carroll, J. A.; and R. Broadhead. *Canadian Internet Handbook.* Toronto: Prentice-Hall Canada Ltd., 1994.

10. Brassard, D. "Information Superhighway." Paper No. BP-385E. Ottawa, Ontario, Canada: Library of Parliament, March 1994.

11. Powell, D. "Supernetworks in Canada Play Catchup." *Computing Canada,* February 1992, p. 6.

12. Pike, M. A., et al. *Using the Internet.* Indianapolis, IN: Que Corporation, 1995.

13. *Using the World Wide Web.* Indianapolis, IN: Que Corporation, 1994.

14. Dieter, G. E. *Engineering Design: A Materials and Processing Approach.* New York: McGraw-Hill, 1983.

15. Lawrence, W. W. *Social Risk Assessment,* ed. R. C. Schwing, and W. A. Albus. New York: Plenum Press, 1980.

16. Abt, C. C. "The Social Role of Technology." In *Technology Assessment and Quality of Life,* ed. G. I. Stober, and D. Schumacher. Amsterdam: Elsevier Scientific Publishing Company, 1973, pp. 31–41.

17. Scheer, L. "Experience with Quality of Life Comparisons." In *The Quality of Life,* ed. A. Szalai, and F. M. Andrews. Beverly Hills, CA: SAGE Publications, Inc., 1980, pp. 145–155.

18. Wilkie, T. "The Unanswered Questions of Chernobyl." *New Scientist,* May 15, 1986, p. 23.

19. Wilkie, T. "Soviet Engineers Admit Failings in Reactor Design." *New Scientist,* August 28, 1986, pp. 14–15.

3

THE DESIGN PROCESS AND IMPORTANT RELATED AREAS

3.1 INTRODUCTION

Producing a marketable product from an initial requirement requires many systematically formulated steps. In other words, the design process may be divided into several phases. The time needed for each phase may vary considerably, and the types of tasks performed may also be quite distinct. The design process itself is the imaginative integration of engineering technology, scientific information, and marketology [1], which is the business of making design-influencing decisions based on consumer reaction, availability of necessary materials, distribution, etc.

Two areas closely related to the design process are: design specification and design review. The main purpose of specifications is to convey, clearly, accurately, the information needed to develop the design. Utmost care is to be exercised during the writing of design specifications.

The number and types of design reviews may vary considerably from one product to another. Design reviews are intended to ensure that the end products are reliable, maintainable, reproducible, faithful to the final design specifications, cost effective, safe, etc. According to some experts [2], the budgeted cost for design reviews should account for somewhere between 1 and 2 percent of the overall engineering cost of a product.

3.2 THE DESIGN PROCESS

Various authors have broken down the design process into several steps, ranging from as few as 5 to as many as 25. For example, G. E. Dieter [3] outlined six steps: requirement recognition, problem definition, information collection, conceptualization, evaluation, and final design communication. In contrast, J. P. Vidosic [4] identified eight steps: need recognition, problem definition, preparation, conceptualization, design synthesis, evaluation, optimization, and presentation, respectively. We will use the 12 stages described by P. H. Hill [2]. These include: problem identification, problem definition, information gathering, task specification, idea generation, conceptualization, analysis, experimentation, solution presentation, production, product distribution, and consumption.

3.2.1 PROBLEM IDENTIFICATION

This stage of the design process calls for rigorous investigation of the problem or requirement. There are various bodies and people that may identify a need, requirement, or problem. Some of these are: customers, marketing agents, government departments and agencies, procurement representatives, trade associations, operators, and servicing people. Generally, a need reflects dissatisfaction with an existing condition, and the desired solution may be expressed in several ways, such as: improve the item's reliability, reduce the cost of the product, or improve product performance and efficiency.

3.2.2 PROBLEM DEFINITION

This second stage of the design process is probably the most crucial of all the other stages and thus, requires the utmost care in establishing problem definition. Past experience indicates that, if conditions permit, it is generally advantageous to have a broad definition of the problem. One reason is that the chances of overlooking unconventional or other solutions to the problem will be significantly reduced, if not completely eliminated. At this stage, the people involved in the design should ask a variety of questions such as:

1. What are the design limitations or constraints?
2. Are there any special difficulties associated with the design?
3. Are there any social consequences of the design?
4. Is the problem under study too big or too complex to handle?
5. What objectives is the design supposed to fulfill?
6. What other factors should be contained in the problem definition?

3.2.3 INFORMATION GATHERING

This stage is concerned with the gathering of appropriate design related information. The sources from which the needed information can be obtained include: professional and trade journals, conference proceedings, textbooks, handbooks, technical reports, catalogs, professional societies and trade associations, government bodies, foreign embassies, suppliers, consumers, and insurance companies.

3.2.4 TASK SPECIFICATIONS

The purpose of task specifications is to help the designers achieve the design objectives. A task specification may simply be described as writing down all important parameters and data tending to control the design and directing it towards the stated goal.

3.2.5 IDEA GENERATION

In this stage, new ideas useful for the ongoing design effort are generated. Designers and others working on the project can take advantage of many techniques developed specifically for this purpose [5], including: checklists, brainstorming [6], synectics, and attribute lists. Some of these techniques are described in detail later in this text.

3.2.6 CONCEPTUALIZATION

This stage of the design process may be regarded as a creative and innovative activity that produces a number of possible alternative solutions to the set goal. The results of conceptualization may take the form of sketches or free-hand drawings.

3.2.7 ANALYSIS

In this important stage of the design process, the proposed alternative solutions are examined against physical laws, the design specifications, and each other, to identify the optimum solution(s).

3.2.8 EXPERIMENTATION

During experimentation, the design is translated into hardware and is tested for performance characteristics, reliability, workability, etc. The three hardware constructs that can be developed at this stage are: prototype, mock-up, and model. A prototype is an actual physical unit built in accordance with the design. Prototypes provide information on such areas as: assembly methods, durability, performance subject to real-life environments, and workability. Although this technique produces the greatest amount of useful information, it is also time-consuming and the most expensive.

The mock-up gives designers a feel for how the design works. Generally, the mock-up is constructed to scale, using materials such as plastic, cardboard, and wood. While the mock-up is the simplest and least expensive to produce, it has one major drawback: it provides the least amount of data. Yet, this approach can be used to sell a new design idea to management and customers, check appearance, review assembly methods, and check clearances.

Models may be classified into four categories: true, adequate, dissimilar, and distorted. A true model meets all the design requirements and is an exact geometric reproduction of the actual physical item. An adequate model is used to test a particular design characteristic and does not normally yield any information with respect to the overall design. A dissimilar model does not resemble the actual physical unit; yet, through appropriate analogies, is quite useful for providing data on behavioral characteristics. The distorted model violates a number of design conditions; yet it fulfills a specific need, such as demonstrating the appearance of an object from an unusual visual perspective.

3.2.9   SOLUTION PRESENTATION

This important stage is concerned with compiling the written report on the design project, so that the design can be easily communicated to all interested parties. The report should include such items as: design description, product operation, needs satisfied by the proposed design, assembly drawings, manufacturing specifications, and standard parts' descriptions.

3.2.10   PRODUCTION

The production stage of the design process is concerned with such areas as: production volume, training, facilities availability, production scheduling, and quality assurance.

3.2.11   PRODUCT DISTRIBUTION

This stage of the design process calls for careful attention to such areas as: product pricing, appropriate product release time, advertising, and appropriate market tests.

3.2.12   CONSUMPTION

This final stage of the design process is concerned with data collection and analysis on such topics as: product performance, reliability, competitors' reactions, and users' reactions. The information thus obtained will provide useful input in the design of new generations of the product.

3.3   DESIGN SPECIFICATIONS

The design specification document provides detailed design-related information to the design engineers and others working on the product. The document serves as a reference and control vehicle when a new product is being designed [7–10]. It is essential that the specifications be given careful consideration when writing design specifications. Guidelines for writing good design specifications are as follows [1]:

1. Use simple, straightforward language.
2. Avoid ambiguous phrases.
3. Minimize references to standard documents.
4. Ensure that the specification is reasonable with respect to set tolerance.
5. Ensure that the specification is accurate and complete.
6. Aim for developing a flexible specification.
7. Avoid repetition.

8. Avoid specifying impossibles.

9. Minimize cross-references.

The design specification serves as a resource document, which may contain information on many different areas, some of which are: objective, product definition, future requirements, special legal requirements, quality requirements (human factors, performance, reliability, appearance, safety, quality assurance, maintenance, storage, and transportation), production levels, product cost, direct and indirect development costs, development timetable, and obsolescence [10].

In many instances, the design specification must be written in a standardized format. Yet, that format may vary from one project to another. In general, the format of a typical design specification is as follows [7–10]:

- Title.
- Date of preparation.
- Author(s) (including name, title, organization, address, etc.).
- Table of Contents.
- Introduction: General background information on the product and the design project.
- Scope: The requirement boundaries and the role of the product, as well as any other relevant information.
- Relevant documents: Other documents relevant to the design specification.
- Actual design specification description: The actual requirements for the product being designed. Generally, this section is broken down into a number of subsections, such as: design and construction, performance, dimensions, materials, maintenance, quality assurance, reliability and maintainability, and acceptance conditions.
- Other required information: Other required documentation, including: test reports, maintenance drawings, operations manuals, etc.
- Appendices.
- Indexes.

There is no doubt that the above section entitled "Actual Design Specification Description" will cover all the specification related material, but in addition, the design specification must include information on such factors as: the environment, installation limitations, manufacturing and operating constraints, functional requirements, and possible effects on other products or systems.

3.4 DESIGN REVIEWS

Product reviews should be conducted throughout the entire life cycle of a product. Of critical importance are the reviews carried out during the product design phase. The design review forms a vital component of modern industrial practice; its basic objective is to insure that appropriate design principles are applied, as well as to determine the progress of the design effort. The objectives of the design reviews can only be achieved if proper care is given to such factors as: the composition of the review team, the appointment of the review board chairman, the issues raised during the reviews, and the frequency of the reviews. According to various experts [2], the cost of such reviews may be between 1 and 2 percent of the overall engineering cost of a product, which means that this cost element should be included in the overall cost estimate of the design project.

3.4.1 DESIGN REVIEW CLASSIFICATIONS

Over the years, different authors, practitioners, and organizations have presented various phases of design reviews. Basically, there are three major review phases:

1. Preliminary design review.
2. Intermediate design review.
3. Final design review.

The preliminary design review is conducted before the initial design is formulated. The main objective of this review is a careful examination of each design specification requirement, from the standpoint of accuracy, validity, and completeness. P. H. Hill [2] and C. L. Carter [11] present a number of items that should be reviewed during this preliminary phase: cost objectives, scheduling requirements, present and future material or component availability, applicable government and other standards, design constraints, design alternatives, customer requirements, applicable legislation, critical components, liability, "make or buy" considerations, required functions, required tests, available data on similar products, and required documentation.

The intermediate design review is conducted prior to the development of the detailed production drawings. At this stage, the design selection process and the preliminary layout drawings should be complete. One of the important tasks of this review is to compare each requirement of the specification with the proposed design. These requirements may concern such areas as: reliability cost, safety, human factors, usage of standard parts, time schedule, maintenance, and performance.

The final design review, which is also called the critical design review, is performed soon after the production drawings are completed. At this stage, the design team should have accumulated a considerable amount of information, such as reports of preceding design reviews, cost data, and test data. Generally, the emphasis of the final design review is on such areas as: design producibility, analysis results, quality control of incoming parts, manufacturing methods, and value engineering.

3.4.2 USEFUL DESIGN REVIEW ITEMS

During the reviews, the design team members use various types of information. This information must be available to all team members. Depending on the nature of the design reviews and the project type and size, the design review committee members usually have access to such items as: a list of parts; design specifications; a list of inputs and outputs; schematic diagrams; test procedures; results of such tests as vibration, thermal, acceleration, and shock; results of failure modes and effect analyses; technical drawings; results of reliability predictions; and reliability analysis techniques.

3.4.3 DESIGN REVIEW COMMITTEE AND BOARD CHAIRMAN

For a fair sized design project, the design review team is comprised of a variety of professionals with the same overall objective, that is, to produce an effective end product. Some of those professionals could be: the design engineer, senior design engineer, component engineer, maintenance engineer, tooling engineer, quality control engineer, test engineer, design review board chairman, customer representatives, packaging and shipping representative, reliability engineer, human factors engineer, and the manufacturing engineer. The number of professionals participating in design reviews may vary from one project to another, and P. H. Hill [2] states there should not be greater than 12 people participating.

The design review board chairman leads the design review team and must always belong to the engineering division. However, the chairperson should not be in a direct line of authority over the designer whose work is being reviewed. The guidelines in Reference 12 state that the configuration manager frequently leads the design review team.

In identifying the design review board chairman, consideration should be given to such factors as: an understanding of the technical problem, personality, degree of tact and discretion, and skill in leading a technical meeting. The design review board chairman performs various tasks, some of which are [1]:

1. Determines the types and frequency of design reviews to be conducted.
2. Chairs the design reviews and establishes guidelines for selecting particular items for review.
3. Circulates the agenda and other materials to everyone concerned well ahead of a scheduled design review.
4. Looks after the publication of the minutes of each review and distributes them to all appropriate bodies.
5. Evaluates the results of each review and directs any necessary follow-up measures.
6. Coordinates with all concerned people and provides assistance if and when desired.

3.4.4 USEFUL LEAD-IN ITEMS AND SUBJECTS FOR CONDUCTING EFFECTIVE DESIGN REVIEWS

To conduct effective design reviews, the board chairman and other professionals participating in the reviews should be aware of areas in which questions can be raised, as well as useful lead-in questions. Some suggested lead-in questions are:

1. Discuss the design procedure to be followed.
2. Identify important requirements associated with the proposed design.
3. Identify expected or already overcome difficulties.
4. Discuss how the proposed design meets the specifications.
5. Provide necessary related background information.

Depending on the specific design project, there are many review areas including: performance, specifications (e.g., validity of specifications, and adherence to specifications), safety, electrical factors (e.g., design simplification, electrical interference, circuit analysis, and testing), standardization, reliability [e.g., results of *failure modes and effect analysis (FMEA)*, redundancy, reliability predictions, reliability allocation, and reliability testing], value engineering, mechanical factors (e.g., thermal analysis, balance, and connectors), drafting (e.g., accuracy and completeness of notes, dimensions, and tolerances), human engineering (e.g., glare, controls and displays, and labeling and marking), maintainability (e.g., maintenance philosophy, parts interchangeability, and minimum downtime), finishing, and final product reproducibility (e.g., assembly economics, parts suppliers, and machine tool effectiveness).

3.4.5 SOFTWARE DESIGN REVIEWS

Today, computers are an important component of engineering systems. The reliability and quality of the computer programs are as important as that of the computer hardware. To ensure this required reliability and quality, software should be subject to design reviews. According to B. W. Boehm [13], about 60 percent of the total software errors are introduced prior to coding. Software design reviews could be used to identify and correct such errors. These reviews may be classified into three categories [14]:

1. *Technical requirements review.* This review is basically concerned with the technical aspects of the software being developed. The review should be conducted soon after documentation of those requirements.
2. *Preliminary design review.* This review is performed after the software is defined to the computer system component level. The review is very useful for evaluating the selected design approach, as well as the functional interfaces.
3. *Critical design review.* This review is concerned with evaluating the fully completed detail design before coding is begun.

3.5 CASE STUDY: HARTFORD CIVIC CENTER COLISEUM DISASTER

The Hartford Civic Center Coliseum roof was one of the largest space frames in the United States. It collapsed on January 18, 1978, as shown in Figure 3.1, because of the weight of snow from a major snowstorm. City government officials established a

Figure. 3.1   Hartford Civic Center Coliseum after the disaster

I UPI/Bettmann

city council panel made up of three members. The panel in turn hired a consulting firm, Lev Zetlin, Associates, Inc. [LZA], of New York City to determine the cause of the disaster.

The report of LZA concluded that design deficiencies led to the collapse of the Coliseum's 2.5–acre steel space frame roof. Specifically, the actual loads on the roof were 1.5 million pounds higher than the designed loads. The report further stated that a computer analysis of the original design indicated that interior and exterior top chord compression members were overloaded by as much as 72 percent and 213–852 percent, respectively. However, a subsequent study of the same disaster attributed the roof collapse to a weld joining the scoreboard to the roof [15].

3.6 PROBLEMS

1. Describe in detail the steps involved in the design process.
2. What are the characteristics of a good design specification?
3. List useful guidelines for writing good design specifications.
4. Describe the general format of a design specification.
5. What are the advantages and disadvantages of design reviews?
6. Discuss the major phases of design reviews.
7. Discuss the important areas which the design review team may question during the review.
8. What are the important functions of the design review board chairman?
9. List the members of a typical design review team.

REFERENCES

1. Dhillon, B.S. *Quality Control, Reliability and Engineering Design.* New York: Marcel Dekker, Inc., 1985.
2. Hill, P.H. *The Science of Engineering Design.* New York: Holt, Rhinehart and Winston, 1970.
3. Dieter, G.E. *Engineering Design.* New York: McGraw-Hill, 1983.
4. Vidosic, J.P. *Elements of Design Engineering.* New York: The Ronald Press Co., 1969.
5. Dhillon, B.S. *Engineering Management.* Lancaster, PA: Technomic Publishing Co., 1987.
6. Osborn, A.F. *Applied Imagination.* New York: Scribner's, 1963.
7. "A Guide to the Preparation of Engineering Specifications," London: The Design Council, 1980.
8. Abbett, R.W. *Engineering Contacts and Specifications.* New York: Wiley, 1967.
9. Pugh, S. *Total Design.* Wokingham, England: Addison-Wesley, 1990.
10. Flurscheim, C.H. "Specifications for the Development of Engineering Products." In *Industrial Design in Engineering,* ed. C.H. Flurscheim. Berlin: Springer-Verlag, 1983, pp. 289–396.
11. Carter, C.L. *The Control and Assurance of Quality, Reliability and Safety.* Richardson, TX: C.L. Carter & Associates, Inc., 1978.
12. *Engineering Design Handbook, Development Guide for Reliability, Part II, Design for Reliability.* AMCP 706–196. Alexandria, VA: U.S. Army Material Command, 1976.
13. Boehm, B.W. "Software Engineering," *IEEE Transactions on Computers* 25, (1976), pp. 1226–1241.
14. McKissick, J. "Quality Control of Computer Software." *Proc. of the Am. Soc. for Quality Control Technical Conf.,* 1977, pp. 391–398.
15. Ross, S.S. *Construction Disasters: Design Failures, Causes, and Prevention.* New York: McGraw-Hill, 1984, pp. 303–329.

4

DESIGN TOOLS

4.1 INTRODUCTION

Design professionals use various design tools, such as optimization methods, project management techniques, idea generation methods, and product marketing techniques. Today's competitive environment demands that a product design be optimized within all possible constraints and variables. At the designer's disposal are a number of optimization approaches to help achieve this objective. Some of these approaches are: differential calculus, Lagrange multipliers, and linear programming.

The management of a design project is as important as the design effort itself. A typical engineering design project may involve the performance of hundreds of tasks for its successful completion. Two effective project management methods widely used in the industrial sector to manage design projects are: the *critical path method (CPM)*, and the *program evaluation and review technique (PERT)*.

In designing, various difficulties can develop, some of which may be overcome without much effort, while others may require innovative solutions. Over the years, researchers involved in developing new idea-generation methods have come up with many creativity techniques. Engineering designers

have traditionally taken advantage of these techniques to find solutions to their problems. One prime example of such methods is the group brainstorming technique; in design work, the professionals concerned have found this to be a very effective tool.

Product marketing information plays an important role in the design effort. Such information may be concerned with the design features preferred by customers, the cost, the background of the end users, and so on. Various ways and means have been developed over the years to obtain relevant product marketing information.

Today, the use of the computer plays a very important role in design work and has led to the development of a process called *computer-aided design*.

This chapter describes all of these items in detail.

4.2 DESIGN OPTIMIZATION METHODS

For design problems, many acceptable solutions may exist, and it is up to the designers to seek the best possible solution by taking into consideration the set objective and the associated constraints.

The process of determining the best solution is known as *optimization*. There are many optimization techniques that give design professionals the capacity to manipulate and refine their work so that the end results are effective and acceptable from all possible directions. Thus, the optimal design may be described as the best of all feasible designs. J.N. Siddall [1] and G. Dieter [2] have categorized design optimization methods into four groups: evolution, intuition, trial and error modeling, and numerical algorithms. This section describes some of the analytical optimization techniques that can be useful in design work.

4.2.1 DIFFERENTIATION METHODS

Several optimization techniques are based on calculus principles. These methods are called the *differentiation methods*. The calculus principles learned by engineers allow them to determine the maximum or minimum value of a differentiable mathematical function. For example, in the case of a multiple-independent-variable function, we take partial derivatives with respect to each of the variables and equate the resulting expressions to zero, as follows:

$$\frac{\partial F(\bullet)}{\partial x_1} = 0 \qquad \textbf{[4.1]}$$

$$\frac{\partial F(\bullet)}{\partial x_2} = 0 \qquad \textbf{[4.2]}$$

$$\frac{\partial F(\bullet)}{\partial x_3} = 0 \qquad \textbf{[4.3]}$$

$$\frac{\partial F(\bullet)}{\partial x_m} = 0 \qquad \textbf{[4.4]}$$

where $F(\bullet)$ is a function of variables $x_1, x_2, x_3, \ldots, x_m$.

Note that a set of m simultaneous equations is yielded by Equations 4.1–4.4. This means that to determine the optimum variable values, these m equations must be solved. For a detailed explanation of this approach, consult a standard textbook on calculus.

This method can be demonstrated for a single-variable function by the following examples [3,5].

Example 4.1 | **A**ssume that the cost of running a motor vehicle is defined by the following equation:

$$K(s) = FC + C_d s^2 \qquad \textbf{[4.5]}$$

where

$K(s)$ is the motor vehicle running cost, expressed in dollars per hour.

FC is the fixed cost associated with the vehicle, even if it is not to be used. (In our case, we assumed this to be $40.00.)

s is the speed of the motor vehicle, expressed in miles per hour.

C_d is the cost associated with the vehicle speed. (In our case, we assumed this to be 0.1.)

Calculate the optimum speed of the motor vehicle when the cost per mile is at its minimum.

Solution

Substituting the given data into Equation 4.5 results in

$$K(s) = 40 + 0.1s^2 \qquad \textbf{[4.6]}$$

In order to obtain the motor vehicle cost per mile, we divide Equation 4.6 by *s* as follows:

$$CM = \frac{K(s)}{s} = \frac{40}{s} + 0.1s \qquad \textbf{[4.7]}$$

where *CM* is the motor vehicle cost per mile.

Obviously, Equation 4.7 is a single-variable differentiable function. Differentiating Equation 4.7 with respect to *s*, we obtain

$$\frac{dCM}{ds} = 0.1 - \frac{40}{s^2} \qquad \textbf{[4.8]}$$

Setting Equation 4.8 equal to zero and rearranging the result leads to

$$s = 20 \text{ miles per hour}$$

Thus, the optimum vehicle speed for the minimum cost per mile is 20 miles per hour.

Assume that the reliability of a mechanical system composed of fluid flow valves is given by | **Example 4.2**

$$RM = (1 - Q_O)^m - Q_c^m \qquad \textbf{[4.9]}$$

where

RM is the mechanical system reliability.

Q_O is the probability of a valve failing in the open mode.

Q_c is the probability of a valve failing in the closed mode.

m is the number of identical, independent redundant valves.

Calculate the number of fluid flow valves needed to obtain maximum system reliability when the failure probabilities of a valve failing in its open mode and closed mode are 0.05 and 0.1, respectively.

Solution

Examining Equation 4.9, we conclude that this is a single-variable differentiable function. Differentiating Equation 4.9 with respect to *m* yields

$$\frac{\partial RM}{\partial m} = (1 - Q_O)^m \, 1n(1 - Q_O) - Q_c^m \, 1nQ_c \qquad \textbf{[4.10]}$$

Setting Equation 4.10 equal to zero and solving for m, we get

$$m* = \frac{\ln\left[\frac{\ln Q_c}{\ln(1 - Q_O)}\right]}{\ln\left[\frac{(1 - Q_O)}{Q_c}\right]} \qquad \textbf{[4.11]}$$

where m^* is the optimum value of m.

Substituting the given data into Equation 4.11 yields

$$m^* = \frac{\ln\left[\frac{\ln 0.1}{\ln(1 - 0.05)}\right]}{\ln\left[\frac{(1 - .05)}{0.1}\right]} = 1.69$$

This result indicates that, to obtain maximum reliability, the mechanical system should have two fluid flow valves.

| **Example 4.3** | **A**ssume that in a manufacturing organization [3], the total cost TC of removing material is expressed by |

$$TC = [(CC)(TL)^{-1} + (OC)](TL)^m / (W\alpha\beta k) \qquad \textbf{[4.12]}$$

where

TC is the total cost of removing material, expressed in dollars per unit.

α is the cut width, expressed in inches.

β is the cut depth, expressed in inches.

TL is the life of the tool, expressed in minutes.

CC is the cost associated with supplying a cutting edge to the machining process.

OC is the cutting cost of a machine operator, including the overhead cost, expressed in dollars per minute.

m, k are the parameters associated with the equation concerning the correlation between the cutting speed and the tool life [$S(TL)^m = k$, where S is the cutting speed at the tool–material interface, expressed in feet per minute]. Both parameters are determined by such factors as: tool material type, tool form and shape, cutting fluid type, material type being machined, cutting tool surface coating, and cut shape and size.

Obtain an expression for the speed required for minimum material removal cost, using the following relationship:

$$S(TL)^m = k \qquad \textbf{[4.13]}$$

Solution

Differentiating Equation 4.12 with respect to TL we get

$$\frac{\partial(TC)}{\partial(TL)} = \frac{1}{W\alpha\beta k} \left[\frac{(CC)(m-1)(TL)^m}{(TL)^2} + \frac{(OC)(m)(TL)^m}{(TL)} \right] \qquad \textbf{[4.14]}$$

Setting Equation 4.14 equal to zero yields

$$\frac{(CC)(m-1)}{(TL)} + (OC)(m) = 0 \qquad \textbf{[4.15]}$$

Then, solving Equation 4.15 for TL yields

$$TL = \frac{(CC)(1-m)}{(OC)m} \qquad \textbf{[4.16]}$$

Substituting Equation 4.16 into Equation 4.13 and then solving for S, we get

$$S^* = k[(CC)(1-m)/(OC)m]^{-m} \qquad \textbf{[4.17]}$$

The expression for the speed required for minimum material removal cost is given by Equation 4.17.

4.2.2 LAGRANGE MULTIPLIER METHOD

This is a powerful approach for determining the optimum values in multivariable problems subject to functional constraints. The method allows the transformation of the objective and constraint functions into a single constraint-free function. For example, if we have a three-variable objective function, $L(x_1, x_2, x_3)$, subject to the functional constraints $N(x_1, x_2, x_3) = 0$, $M(x_1, x_2, x_3) = 0$, and $W(x_1, x_2, x_3) = 0$, the constraint-free function $Q(x_1, x_2, x_3)$ will be

$$\begin{aligned} Q(x_1, x_2, x_3) =& L(x_1, x_2, x_3) + \lambda_1 N(x_1, x_2, x_3) \\ &+ \lambda_2 M(x_1, x_2, x_3) + \lambda_3 W(x_1, x_2, x_3) \end{aligned} \qquad \textbf{[4.18]}$$

where λ_1, λ_2, and λ_3 are known as the *Lagrange multipliers.*

For the optimum solution, the following conditions must exist:

$$\frac{\partial Q}{\partial x_1} = 0 \qquad \textbf{[4.19]}$$

$$\frac{\partial Q}{\partial x_2} = 0 \qquad \textbf{[4.20]}$$

$$\frac{\partial Q}{\partial x_3} = 0 \qquad \textbf{[4.21]}$$

$$\frac{\partial Q}{\partial \lambda_1} = 0 \qquad \textbf{[4.22]}$$

$$\frac{\partial Q}{\partial \lambda_2} = 0 \qquad \textbf{[4.23]}$$

$$\frac{\partial Q}{\partial \lambda_3} = 0 \qquad \textbf{[4.24]}$$

The optimum values of x_1, x_2, and x_3 are obtained by solving Equations 4.19–4.24. A problem with m variables and n constraints may be formulated and solved in a similar manner,

---

Example 4.4 | **A**ssume that to have a satisfactory heat-transfer surface area in a heat exchanger, a total of 450 linear feet of tubing is required [6]. The cost k of the installation, expressed in dollars, is given by

$$K = SC + TC + FSC \qquad \textbf{[4.25]}$$

where

SC   is the cost associated with the shell.

TC   is the cost associated with the tubes.

FSC is the cost associated with the floor space required by the heat exchanger.

The three costs of Equation 4.25 are defined as follows:

$$SC = 40md^{2.5} \qquad \textbf{[4.26]}$$

where d is the heat exchanger diameter and m is the heat exchanger length.

$$TC = \$950 \qquad \textbf{[4.27]}$$

$$FSC = 40md \qquad \textbf{[4.28]}$$

If Equation 4.25 is subject to the following constraint,

$$m(40)\frac{\pi d^2}{4} = 450 \qquad \textbf{[4.29]}$$

determine the values of m and d for a minimum K by assuming that the tube spacing will allow 40 tubes to fit in a cross-sectional area of 1 ft^2 in the interior of the shell [2].

Solution

Rearranging Equation 4.29 yields

$$m - \frac{45}{\pi d^2} = 0 \qquad \textbf{[4.30]}$$

Thus, the constraint-free Lagrangian function, using Equations 4.25–4.28 and 4.30, is

$$Q = 40md^{2.5} + 950 + 40md + \lambda\left(m - \frac{45}{\pi d^2}\right) \qquad \textbf{[4.31]}$$

Differentiating Equation 4.31 with respect to m, d, and λ and then equating the resulting expressions to zero yields

$$\frac{\partial Q}{\partial m} = 40d^{2.5} + 40d + \lambda = 0 \qquad\qquad \textbf{[4.32]}$$

$$\frac{\partial Q}{\partial d} = 100md^{1.5} + 40m + \frac{90\lambda}{\pi d^3} = 0 \qquad\qquad \textbf{[4.33]}$$

$$\frac{\partial Q}{\partial \lambda} = m - \frac{45}{md^2} = 0 \qquad\qquad \textbf{[4.34]}$$

Solving Equations 4.32–4.34, we get

$$d = 1.59 \text{ ft} \qquad \text{and} \qquad m = 5.68 \text{ ft}$$

Thus, the values of d and m for the minimum cost K are 1.59 ft and 5.68 ft, respectively.

4.2.3 LINEAR PROGRAMMING METHODS

Linear programming is a very powerful technique that can be used to optimize various design problems under given constraints. Developed by George B. Dantzig, linear programming was initially known as "programming of independent activities in a linear structure" [7]. In 1947, Mr. Dantzig developed the simplex method of linear programming for the United States Air Force, and the early applications were concerned with military problems, such as procurement, logistics, and transportation. Since 1947, many researchers, including W.W. Cooper, W. Orchard-Hays, A. Charnes, and A. Henderson, have explored different applications of this technique. The rapid developments in computer technology have also helped to increase the applications of linear programming to many other areas, such as: agriculture, business, industry, and space exploration.

In simple terms, linear programming may be described as a mathematical method of allocating constrained resources to attain an objective, such as to minimize cost or maximize profit. Examples of the resources referred to are: time, labor, material, and money. Linear programming may also be said to involve the description of a real-life problem as a mathematical model consisting of linear constraints and a linear objective function.

The three basic steps involved in developing a linear programming mathematical model are as follows [8]:

1. Define the associated decision variables.
2. Define the associated objective function.
3. Define the associated constraints.

The generalized version of the linear programming mathematical model may be expressed as:

Minimize (or maximize) $M = c_1 y_1 + c_2 y_2 + c_3 y_3 + \ldots + c_m y_m$ (objective function)
$$\textbf{[4.35]}$$

subject to the constraints

$$b_{11}y_1 + b_{12}y_2 + b_{13}y_3 + \cdots + b_{1m}y_m (\leq, =, \geq) R_1$$
$$b_{21}y_1 + b_{22}y_2 + b_{23}y_3 + \cdots + b_{2m}y_m (\leq, =, \geq) R_2$$
$$b_{31}y_1 + b_{32}y_2 + b_{33}y_3 + \cdots + b_{3m}y_m (\leq, =, \geq) R_3 \qquad \textbf{[4.36]}$$
$$\vdots \qquad \vdots \qquad \vdots \qquad \qquad \vdots \qquad \vdots$$
$$b_{k1}y_1 + b_{k2}y_2 + b_{k3}y_3 + \cdots + b_{km}y_m (\leq, =, \geq) R_K$$

$$y_1, y_2, y_3, \cdots, y_m \geq 0 \qquad \textbf{[4.37]}$$

where

$m =$ the total number of decision variables.

$k =$ the total number of constraints.

$y_i =$ the ith decision variable, for $i = 1, 2, 3, 4, \ldots, m$.

$b_{ji} =$ the amount of resource the decision variable y_i consumed per unit of activity, for $j = 1, 2, 3, \ldots, k$ and $i = 1, 2, 3, \ldots, m$.

$R_j =$ the total resource, for $j = 1, 2, 3, \ldots, k$.

$c_i =$ the penalty (contribution) per unit of activity for $i = 1, 2, 3, \ldots, m$.

Alternatively, Equations 4.35–4.37 may be expressed as:

$$Minimize\ (or\ maximize)\ M = \sum_{i=1}^{M} c_i y_i \qquad \textbf{[4.38]}$$

subject to the constraints

$$\sum_{i=1}^{m} b_{ji} y_i (\leq, =, \geq) R_j, \qquad for\ j = 1, 2, 3, \ldots, k \qquad \textbf{[4.39]}$$

$$y_i \geq 0, \qquad for\ i = 1, 2, 3, \ldots, m \qquad \textbf{[4.40]}$$

A linear programming problem may be solved either analytically or graphically. However, the graphical approach, which is generally quite effective for two decision variables, would be difficult to use for more than two such variables. For illustration purposes, we will only present the graphical method. The details of the analytical procedure may be found in References 7 and 8.

Example 4.5 | **A**ssume that a department in a small manufacturing organization produces two types of parts, A and B. Further assume that they produce y_1 units of type A and y_2 units of type B. Each part is manufactured by a two-step process involving two robots, M and N. The manufacturing times for y_1 units of A on robots M and N are four hours and six hours, respectively. Similarly, the manufacturing times for y_2 units of B on robots M and N are eight hours and four hours, respectively.

For the upcoming three-week period, robot M will be available for 120 hours and robot N for 100 hours. It is estimated that each manufactured unit of type A generates a profit of $40 and type B generates $30. The organization is in position to sell the parts as soon as they are manufactured. Estimate graphically how many units of each part type should be manufactured to maximize profit.

Solution

In this case, the objective function is the profit function which is expressed as

$$Z = 40y_1 + 30y_2 \qquad \text{[4.41]}$$

where Z is the total profit.

The availability time constraints for robots M and N respectively are

$$4y_1 + 8y_2 \leq 120 \qquad \text{[4.42]}$$

$$6y_1 + 4y_2 \leq 100 \qquad \text{[4.43]}$$

The minimum production for each part type is zero. Thus, we also have

$$y_1 \geq 0 \quad and \quad y_2 \geq 0$$

Rewriting this problem in a linear programming form, we have:
Maximize

$$Z = 40y_1 + 30y_2 \qquad \text{[4.44]}$$

Subject to

$$4y_1 + 8y_2 \leq 120 \qquad \text{[4.45]}$$

$$6y_1 + 4y_2 \leq 100 \qquad \text{[4.46]}$$

$$y_1 \geq 0 \qquad \text{[4.47]}$$

$$y_2 \geq 0 \qquad \text{[4.48]}$$

Figure 4.1 shows the plots of Equations 4.44–4.48. The plot of Equation 4.44 for $Z = 600$ indicates that the optimum profit is greater than $600. On the other hand, the plot of the same equation for $Z = \$1200$ demonstrates that this value of Z does not satisfy all the constraints; thus, the optimum profit is much lower. As shown in the figure, the optimum profit is generated at point P. At this point, all the constraints are fully satisfied. The corresponding values of y_1 and y_2, in this case, are equal to 10. Substituting these values into Equation 4.44 yields the optimum profit of $700.

4.3 PROJECT MANAGEMENT METHODS

The management of engineering design projects is as important as the design effort itself. Over the years several methods and procedures have been proposed to manage

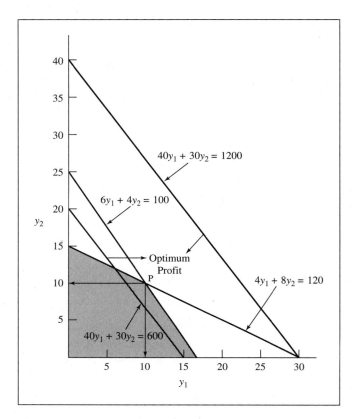

Figure 4.1   Plots of constraints and the objective function

such projects effectively. The critical path method (CPM) and the program evaluation and review technique (PERT) [11], which were originally developed during the same time period (i.e. the late 1950s), are the two most widely used approaches. Today, these two methods are computerized.

PERT was developed under the guidance of the U.S. Navy Special Projects Office by a team that included members from the Lockheed Missile Systems Division and the consulting firm Booz, Allen and Hamilton [9,10]. The technique was developed to monitor the effort of 250 main contractors and 9,000 subcontractors concerned with the Polaris missile project. CPM was the result of the efforts of the Integrated Engineering Control Group of E.I. duPont de Nemours & Company, and was developed to monitor activities related to design and construction. Later, DuPont and Remington Rand Corporation implemented CPM on a computer and used the program to construct a $10 million chemical plant in Louisville, Kentucky, in 1957. The CPM network contained over 800 activities, and the technique was managed by six engineering professionals.

A survey of manufacturing companies in the United States revealed that among the operations research techniques used, CPM and PERT accounted for over 65 percent [12]. Some of the areas in which CPM and PERT are used are: construction, research and development, product planning, maintenance, and computer installation.

A CPM and PERT project generally has the following characteristics [13]:

1. There are clearly defined activities or jobs whose accomplishment results in project completion.

2. Once started, the activity or job continues without interruption until the end.

3. The activities or tasks are independent, which means they may be started, stopped, and performed individually in a prescribed sequence.

4. The activities or jobs are ordered and they follow each other in a specified manner.

Even though the CPM and PERT techniques were developed independently, the basic theory and symbology in both the techniques are essentially the same. In general, CPM is used in situations where the activity duration times are reasonably predictable with a good degree of accuracy, such as in the construction industry. In contrast, PERT is used in situations where the activity duration time estimates are generally uncertain, such as in research and development projects. The steps involved in both methods are presented here.

4.3.1 SYMBOLS, TERMS, AND DEFINITIONS

Several symbols, terms, and definitions are used in developing the CPM and PERT networks. The basic symbols and terms for both systems are essentially the same [13,14] and are as follows:

1. **Event (Node).** This represents a point in time in the life of a project. An event can be the beginning or the end of an activity. A circle [O] is used to represent an event. Generally, each network event is identified with a number.

2. **Activity.** This is an effort needed to carry out a certain portion of the project. An activity consumes money, time, labor and materials, and is denoted by a continuous arrow [→]. The direction of progress on the project is indicated by the arrowhead. Further, an activity both begins and ends at a circle.

3. **Dummy activity.** This is an imaginary activity that takes zero time and no resources to perform. Its purpose is to show appropriate relationships between activities. A dotted arrow [---→] is used to represent such an activity.

4. **Network paths.** These are the paths used or needed to reach the project termination point or event (i.e., the final event) from the project starting point or event.

5. **Critical path.** This is the longest path (with respect to time duration) through the PERT or CPM network. In other words, the critical path is the largest sum of activity durations of all the individual network paths. If the completion of any activity

on the critical path is delayed, then the predicted completion date for the entire project will be delayed.

6. **Earliest event time (EET).** This is the earliest time at which an event occurs, provided all preceding activities are accomplished within their estimated times.

7. **Latest event time (LET).** This is the latest time at which an event could be reached without delaying the predicted project completion date.

8. **Total float.** This is the latest time of an event minus the earliest time of the preceding event and the duration time of the in-between activity.

Several points are useful for constructing a CPM or PERT network. These are:

1. An activity begins at a node (event) and terminates at an event.
2. In general, an activity can only be started after reaching its tail event.
3. It is impossible to reach an event until all activities leading to it are accomplished.

4.3.2 CPM AND PERT NETWORK DEVELOPMENT

Steps involved in constructing a CPM network are as follows [13]:

1. Break down the design (or other) project into individual activities and identify each activity.
2. Estimate the time required for each activity.
3. Determine the activity sequence.
4. Construct the CPM network, using the defined symbols.
5. Determine the critical path of the network.

Similarly, the steps involved in developing a PERT network are [12]:

1. Break down the design (or other) project into individual activities and identify each activity.
2. Determine the activity sequence.
3. Develop the PERT network, using the defined symbols.
4. Obtain the expected time to perform each activity, using the following formula [15]:

$$T_e = \frac{x + 4y + z}{6} \qquad [4.49]$$

where

T_e is expected time for the activity.

x is the optimistic time estimate for the activity.

y is the most likely time estimate for the activity.

z is the pessimistic time estimate for the activity.

5. Determine the network critical path.
6. Compute the variance (σ^2) associated with the estimated expected time T_e of each activity using the following formula:

$$\sigma^2 = \left(\frac{z-x}{6}\right)^2 \qquad \text{[4.50]}$$

The symbols x and z were defined earlier.

7. Obtain the probability of accomplishing the design (or other) project on the stated date, using the following formula:

$$w = \frac{T - T_L}{\left[\sum \sigma_{cr}^2\right]^{1/2}} \qquad \text{[4.51]}$$

where

σ_{cr}^2 is the variance of activities on the critical path.

T_L is the last activity's earliest expected completion time, as calculated through the network.

T is the design project due date, expressed in time units such as days, weeks, or years.

Table 4.1 presents the probabilities for selective values of w. To obtain the probability of completing the design (or other) project on a specified date, calculate the value of w using Equation 4.51 and obtain the corresponding probability value from Table 4.1.

Example 4.6

A mechanical design project was broken down into a number of major jobs or activities, as shown in Table 4.2. Develop a CPM network using the data given in the table, and determine the critical path of the network, along with the expected project duration time period.

Solution

Using the defined symbols and the data given in Table 4.2, the CPM network in Figure 4.2 was developed. The paths originating at event 1 and terminating at event 11 are as follows:

1. A-B-C-D-E-G-I-J
2. A-B-C-D-F-G-I-J
3. A-B-C-D-E-G-H-J
4. A-B-C-D-F-G-H-J

The activity duration times for each path are summed, with the following results:

1. $1+2+1+4+3+2+4+3 = 20$
2. $1+2+1+4+4+2+4+3 = 21$ (longest or critical path)

Table 4.1 Standardized normal distribution function approximate values [1]

w	Probability
−3	0.0013
−2.5	0.006
−2	0.023
−1.5	0.067
−1	0.159
−0.5	0.309
0	0.5
0.5	0.69
1	0.84
1.5	0.933
2	0.977
2.5	0.994
3	0.999

[1]For more values, consult the standard normal distribution tables given in a standard probability and statistics book.

Table 4.2 A mechanical design project activity breakdown and duration estimates

Activity Description	Activity Identification	Immediate Predecessor Activity or Activities	Activity Duration (week or weeks)
Group formation	A	—	1
Literature search and design idea generation	B	A	2
Idea evaluation and selection	C	B	1
Design and evaluation	D	C	4
Part manufacture	E	D	3
Part procurement	F	D	4
Design assembly	G	E,F	2
Testing	H	G	2
Cost estimation	I	G	4
Report writing and presentation preparation	J	H,I	3

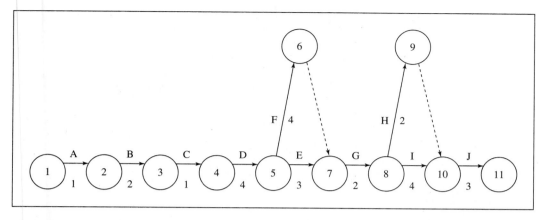

Figure 4.2   CPM network for Table 4.2 data

3. $1 + 2 + 1 + 4 + 3 + 2 + 2 + 3 = 18$
4. $1 + 2 + 1 + 4 + 4 + 2 + 2 + 3 = 19$

These results indicate that the critical, longest, path is represented by activities A-B-C-D-F-G-I-J, and the predicted completion time for the design project is 21 weeks.

Example 4.7

Assume that for Example 4.6 the optimistic, most likely, and pessimistic activity times are as given in Table 4.3. Calculate each activity expected time and variance and the probability of accomplishing the design project in 32.5 weeks. In addition, calculate each event's earliest and latest event times.

Solution

Using the data given in Table 4.3 in Equations 4.49 and 4.50, we obtained expected time and variance for each activity, as shown in the last two columns of Table 4.3.

Before going further, we will first describe the conventions associated with the event symbol. Generally, the symbol used to represent an event is divided into three parts:

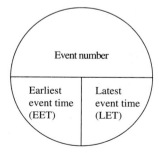

Table 4.3   **A**ctivity time estimates and variances

Activity Identification	Activity Time Estimates (weeks)			Activity Expected Time (T_e)(weeks)	Activity Variance (σ^2)
	Pessimistic (z)	Most Likely (y)	Optimistic (x)		
A (1–2)	4	2	1	2.16	0.25
B (2–3)	3	2	1	2	0.11
C (3–4)	4	2	1	2.16	0.25
D (4–5)	8	5	3	5.16	0.7
E (5–7)	5	4	2	3.83	0.25
F (5–6)	7	4	3	4.33	0.44
G (7–8)	5	3	2	3.17	0.25
H (8–9)	5	3	2	3.17	0.25
I (8–10)	9	6	3	6	1
J (10–11)	7	4	2	4.17	0.69

Thus, for a two-event network, we would have

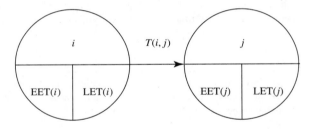

The symbols used in the diagram are defined as follows:

i ith event.

j jth event.

EET(i)   earliest event time of event i.

LET(i)   latest event time of event i.

EET(j)   earliest event time of event j.

LET(j)   latest event time of event j.

T(i,j)   duration time of the activity between events i and j.

The earliest event time of event j can be calculated by following a forward pass and using the following relationship:

$$EET\,(j) = \text{maximum for all preceding events i of}\,[\,EET(i) + T(i,j)\,] \qquad \textbf{[4.52]}$$

Similarly, the latest event time of event i can be computed by following a backward pass and using the following relationship:

LET(i) = minimum for all succeeding events j of $[LET(j) - T(i,j)]$ **[4.53]**

Note that for the network's first and last events the earliest and latest event times, respectively, are equal and zero. Redrawing the network of Figure 4.2, using the new symbol to represent an event and the calculated values of activity expected time and variance from Table 4.3, we obtain the network shown in Figure 4.3. The earliest and latest event times for each event were calculated using Equations 4.52 and 4.53. The earliest expected completion time for the last activity (10–11) is 29.15 weeks, as shown in the figure. The critical path of the net-

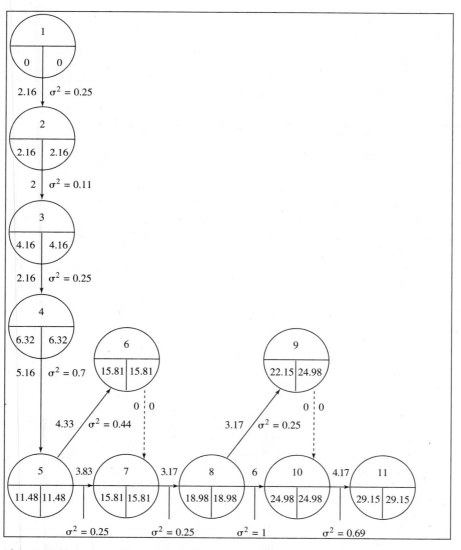

Figure 4.3   Network using the new event symbol and the data in Table 4.3

work is 1–2, 2–3, 3–4, 4–5, 5–6, 6–7, 7–8, 8–9, 9–10, and 10–11. Thus, the total variance of activities on the critical path is given by

$$\sigma_{TC}^2 = \sigma_{1-2}^2 + \sigma_{2-3}^2 + \sigma_{3-4}^2 + \sigma_{4-5}^2 + \sigma_{5-6}^2$$
$$+ \sigma_{6-7}^2 + \sigma_{7-8}^2 + \sigma_{8-10}^2 + \sigma_{10-11}^2$$
$$\sigma_{TC}^2 = (0.25) + (0.11) + (0.25) + (0.7) + (0.44)$$
$$+ 0 + (0.25) + 1 + (0.69) = 2.69$$

Substituting the given and calculated values into Equation 4.51 yields

$$w = \frac{32.5 - 29.15}{(2.69)^{1/2}} \approx 2$$

Using Table 4.1, a w value of 2 yields a probability of 0.997. This means that the probability of completing the design project in 32.5 weeks is 99.74 percent.

4.3.3 ALTERNATIVE NETWORK CRITICAL PATH APPROACH

An alternative approach to determining the critical path is as follows:

1. Compute the earliest and latest event times for each network event.

2. Identify those network events whose earliest and latest event times are equal. If these events are along a single path from the first network event to the last network event, this path is the network critical path. If, however, the events are in more than one path, proceed to the next step.

3. Calculate the total float of each activity for all the paths identified in Step (2). Sum the total floats of each path. The path with the least sum of the total floats is the network's critical path. The total float for an activity, such as between events i and j, is given by

$$\text{Total float} = LET(j) - EET(i) - T(i,j) \qquad \textbf{[4.54]}$$

4.4 DESIGN CREATIVITY

In design work, from time to time concerned professionals are faced with finding solutions to various types of problems. Creative thinking often plays a crucial role. If we look back to some of the important past discoveries, such as the automobile, wheel, television, and telephone, all of which have shaped our daily lives, we see that they were the result of creative thinking. According to Rothenberg and Greenberg [16], over 8,000 publications on creativity appeared during the period 1566–1774.

As the complexity of engineering devices has grown tremendously over the last two centuries, the designing of a product has evolved beyond being a single person's task. For example, in 1840, a Springfield rifle had 140 parts; in 1980, the Boeing 747 jumbo aircraft contained over five million parts and used over 10,000 man-years of design time of thousands of design professionals [17]. In situations such as these, the concern is not only with the creativity of an individual, but with that of an entire group. This section discusses the various aspects of creativity.

4.4.1 CREATIVE PROBLEM SOLVING AND GUIDELINES FOR ENHANCING CREATIVE THINKING

There are basically six steps involved in creative problem solving [18] as follows:

1. Identify the problem and establish its definitions.
2. Collect and analyze relevant data.
3. Generate new ideas and alternatives.
4. Select the most appropriate idea or ideas.
5. Verify the proposed solution or solutions.
6. Implement the resulting final solution.

Over the years, many people have presented various guidelines for improving creative thinking [19,20]. Some of the important ones are as follows [2,21].

1. Develop an attitude toward creative thinking.
2. Unlatch your personal imagination using methods presented in published literature, such as asking the questions "why" and "what if".
3. Develop persistence. Edison tested over 6000 materials prior to his successful discovery of a bamboo species as a filament for the incandescent light bulb. He stated that an invention is 5 percent inspiration and 95 percent perspiration.
4. Develop a receptive attitude towards new ideas, regardless of their origin.
5. Do not pass critical judgment on an emerging idea.
6. Develop boundaries of the problem in question.

4.4.2 CREATIVITY PROMOTION AND PREVENTION FACTORS

Creativity is instrumental in successful design projects. It is therefore essential to be aware of the personal barriers to creativity, as well as the factors that can promote creativity. Examples of personal barriers are: fixed or inflexible thought processes, habits, premature judgment, fear of criticism, overspecialization, dependence on authority, fear of failure, and inadequate background preparation [22,23]. In contrast, some of the factors that enhance creativity are: broad knowledge and interest; freedom from repetitive work habits and fixed mental processes, review of earlier work, good powers of observation and memory, creative mind, interest in abstract thinking, and ability to use analogous conditions [22–24].

4.4.3 CREATIVITY METHODS

Over the years, many idea generation methods have been developed. Their application and complexity may vary from one method to another. According to some who have reviewed these methods, there are over 30 techniques that could be employed to find solutions to various problems. Eight of these methods are described in Reference 24 and some important ones are presented here.

Checklist This is one of the simplest and most useful methods for producing solutions to a problem [25]. The method calls for preparing a list of general questions on the problem, and then seeking answers to those questions. However, the procedure makes the assumption that an idea or solution exists. Some typical questions are as follows [3, 6, 25]:

1. Can the present advantages of the existing solution be further improved?
2. Are there any other applications of the existing solution?
3. Can the existing solution be modified?
4. Can the drawbacks to the existing solution be overcome?
5. How can the performance, quality, and appearance of the existing solution be further improved?
6. Are there any other scientific bases equally effective for the solution in question?
7. Can the solution be combined with another solution to make it more effective?
8. Is it possible to enlarge or reduce the solution or feature?
9. What are the consequences if the solution is taken to its extreme?
10. Is it possible to rearrange parts?
11. Is it possible to make the solution more compact?

Gordon Method This method was developed by W.J. Gordon [26] and successfully applied to the development of new types of products, such as a can opener, a razor, and a gasoline pump. The technique is based on a group effort, and generally needs on average of one day of discussion. Six people formed Gordon's team, and his method possessed two important characteristics:

1. The first characteristic is that the team or group explores the underlying concept of the problem. For example, in developing a new can opener, the group would examine the concept of opening.

2. The second characteristic is that the group investigate the subject from a number of different aspects.

In the case of the can opener, the team members would first study the meanings of the word "opening" and would then study examples of opening. This is useful for uncovering unconventional approaches. Once these approaches are identified, the team then develops the approaches.

The two advantages of this method are: (1) they encourage the radical application of techniques already developed, and (2) they prevent early-stage closure on the problem being addressed.

CNB Method   This is another team-based creativity method. It assumes that the members of the group understand the problem objective and are receptive to cooperation. The following steps are associated with this approach [27]:

1. Distribute to each participating individual a package containing the problem description, a notebook, relevant creative aids and preparational material.

2. Allow each person a specified amount of time, such as a month, to solve the problem, and request that all ideas be recorded each day of that period. At the end of the period, each person chooses the most fruitful idea, summarizes the remaining ideas, and documents their thoughts in the notebook for further exploration.

3. Collect the notebooks, study the proposed ideas, and prepare a detailed report.

4. Invite all team members to review all the notebooks.

5. Schedule a meeting of all participating individuals to discuss the proposed solutions and select the most promising solution.

Group Brainstorming Method   This technique for generating ideas is perhaps the most widely used. It is based on the effort of individuals working in a group environment. The method was first applied on the modern line by Alex Osborn, the author of *Applied Imagination* [18]. The basis for the technique is that, during a group session, one proposed idea for the problem may trigger other ideas (usually, in another individual's mind) and the process continues. The published literature over the years has indicated that the best results are obtained when 8–12 people with different backgrounds but similar interests participate in each brainstorming session. The length of time spent on each brainstorming session is always less than one hour, and sometimes as short as 15 minutes; however the concentration on a given problem is very intense. Additional guidelines associated with this approach are as follows:

1. Strive for quantity of ideas during each brainstorming session. The rule of thumb is to aim for at least 50 ideas, right or wrong, during each session. Remember that it is easier to delete ideas than to think them up. Welcome free-wheeling discussions.

2. Allow absolutely no criticism of any idea during the session.

3. Keep the ranks of participating individuals fairly equal; otherwise, it may take a lot of warm-up time for low-ranking participatants to mix their ideas freely with those generated by more senior members.

4. Devise an effective means for recording the ideas presented. In the past, experience has shown a blackboard is a very effective tool for this very purpose. One clearcut advantage of the blackboard is that every participating member can visualize the ideas generated.

5. During recording, aim at keeping participants' responses as short as possible.

6. Prepare some relevant questions ahead of time, to stimulate ideas in case participant output lags.

7. Carefully select a brainstorming session leader (if applicable).

8. Provide participating members with necessary information on the forthcoming session well in advance, so they have sufficient time to prepare.

9. Appoint a board (as necessary) to screen out the best idea from those proposed during the session, and to combine and improve ideas (if appropriate).

Example 4.8 | **A** brainstorming session was conducted at an American Society of Training Directors workshop. The brainstorming problem was "How to Kill Ideas" The session generated a total of 56 idea chillers. Some of those were:

1. It will cost too much.
2. It is not our problem.
3. We have not budgeted for it.
4. It will be too difficult to sell.
5. There is not enough time.
6. Our organization is too small for it
7. We have already tried that before
8. It is not practical.
9. Others will laugh at us.
10. Let's establish a committee to study it.
11. The union will not be happy.
12. Your idea is too radical.
13. We have no expertise.
14. This is not in our jurisdiction
15. It will be a difficult sell to the top management.

4.5 PRODUCT MARKETING

In general, a designed product can only be considered successful if it is well received by its intended market. Therefore, it is absolutely vital to consider the marketing aspect of the product at the initial design phase. Professionals involved in design must clearly understand the importance of marketing. Conceptually, marketing is about achieving "the right product (P) at the right price (P) with the right promotion (P) in the right place (P)." [28] This is known as the 4Ps concept.

4.5.1 MARKETING FUNCTIONS

Various authors have identified a varied number of marketing functions, from 7 [29] to 47 [31]. For our purposes, we have divided these functions into seven categories [1], as follows:

1. **Advertising.** This is basically concerned with informing end users of the availability of goods and services. Tasks associated with this function include: preparing the advertising message, determining the budget, selecting an advertisement agency, and testing the advertisements for effectiveness.

2. **Pricing.** This is concerned with determining the proper price for goods and services. Decisions are made on topics such as: basic price setting approach, discount policy, price objective, and actual price.

3. **Gathering market information.** Generally, the marketing department provides market-related information to other groups in the organization. In order to perform such a task effectively, the marketing group must collect the required information from various sources. The types of information collected include: population size and trends in a given area, sales volumes, company image, expected competition, type and expected numbers of potential customers, and type of advertising needed.

4. **Personal selling.** This function uses personal resources to inform potential customers about the availability of goods and services. In this case, the salesperson actually calls on potential customers. Tasks involved in this function include: choosing the potential customers to call; determining the optimum period for such calls; selecting, hiring, and training appropriate salespersons; supervising, and conducting performance evaluations of such salespersons; and motivating and compensating the salespersons selected.

5. **Product distribution.** This is concerned with moving the manufactured goods from the manufacturing plant to the ultimate users. Two major components of this function are: transportation, and storage facilities. In this case, decisions are made on such items as: the type, size, number, and location of warehouses; mode of transportation; shipment size, customer service needs; and optimum number of items in the inventory.

6. **Goods and services management.** This function is usually conducted in conjunction with such departments as: research and development, and manufacturing. One important task of this function deals with meeting the requirements of the customer. Each product to be marketed requires decisions on such items as: style, packaging, size, raw materials, durability, warranty, and color.

7. **Distribution channel management.** This is basically concerned with "middle people," such as wholesalers and retailers, which are generally used by manufacturing companies to market their products. Some of the areas involved in this function are: deciding whether or not to use middle people, selecting the middle people, and analyzing the effectiveness of the distribution channels.

4.5.2   A MARKETING PROCEDURE

There are several ways to approach the marketing function. One such approach is composed of the following seven steps [29]:

1. **Conduct market analysis.** This step is basically concerned with identifying market needs, and then ascertaining the type of market to be served and the best way to serve it, and so on.

2. **Conduct environmental analysis.** This step deals with performing analyses of such items as: the competition, legal aspects, culture, technology, and economy.

3. **Develop feasible objectives.** This step is concerned with establishing objectives both strategic and tactical. The strategic objectives represent the primary objectives of the company, and usually take over a year to complete. Strategic objectives would include: profit margin, market share, growth, and sales volume. Tactical objectives are also accomplished in a year and are generally assigned to a company's marketing and other departments.

4. **Accomplish strategic objectives through product/market combinations.** In this step, the marketing department and management evaluate various combinations of product and market to determine how to accomplish the strategic objectives. The four product–market combinations are as follows:

   a.   Same product and same market.

   b.   Same product and different market.

   c.   Different product and same market.

   d.   Different product and different market.

5. **Accomplish tactical objectives through marketing mix.** The term *marketing mix* implies the use of various marketing decision combinations to sell goods to specific markets over a specified period. Thus, the tactical objectives are achieved through the correct marketing mix. Two examples of areas where improvements may be required are: efficiency of a salesperson's traveling times and changes in the advertising medium to increase the customer population.

6. **Accomplish objectives through marketing organization.** This step is basically concerned with achieving the set objectives by developing an effective marketing organization. Areas of concern at this stage include: motivation of salespeople, span of control, effective organizational setup, and so on.

7. **Develop a feedback control system.** This step is concerned with monitoring the performance of the marketing department. The step therefore calls for developing: performance standards, performance measurement criteria, etc.

4.5.3 MARKETING CHECKLIST AND POOR PRODUCT PERFORMANCE

A marketing department must collect various types of information, which may be classified into three broad categories: market opportunity, distribution, and competition. These categories are in turn comprised of several types of information, as follows:

1. Market opportunity.
   - Potential customers (number and type).
   - Potential customers' location.
   - Potential sales volume (in units and dollars).
   - Potential customers' buying habits.
   - Customers' reasons for buying the proposed product.
   - Market growth.
   - Price range of the proposed product.

2. Distribution.
   - Distribution channels, including policies important to them and their ways of conducting business.
   - The company's image in industry.
   - Promotional effort needs.
   - Product compatibility with established channels.

3. Competition.
   - Expected competition.
   - Competitors' image and selling policies.
   - Success level needed to compete with existing products.
   - Time required to improve the existing product.
   - Ease with which product can be copied by others.
   - Unforeseen government control.

Over the years, various studies have been conducted to identify the causes of products unsuccessful in the marketplace. For example, according to Diamond and Pintel [30], 80 percent of the 9,450 new supermarket products were unsuccessful. Different investigations have pointed out various causes of unsuccessful products. Some of these causes are:

1. Poor reliability and/or quality.
2. Too high a selling price.
3. Bad timing for the introduction.

4. Poor performance.

5. Poor marketing strategies.

6. The wrong market.

7. Insignificant difference between new product and successful competing products.

8. Failure to conduct at least limited market tests prior to product release.

4.5.4 SELECTIVE MATHEMATICAL MODELS FOR MAKING MARKETING-RELATED DECISIONS

Published literature on the subject of marketing contains many mathematical models useful for making decisions. In this section, we discuss two of these models. Several others are given in Reference 10.

Buying-Power Index Model This model is an index used for predicting the percentage of total national buying-power for a given area. The index is defined by [32]

$$BP = 0.2L + 0.5M + 0.3N \qquad \textbf{[4.55]}$$

where

BP is the percentage of the total national buying-power.

N is the percentage of the total national retail sales in a given area.

L is the percentage of the total national population residing in a given area.

M is the percentage of the total national disposable personal income in a given area.

Example 4.9

Assume that for a given city, the values of L, M, and N are as follows: $L = 0.15$ percent, $M = 0.6$ percent, and $N = 0.3$ percent.

Compute the value of the buying-power index for that city.

Solution

Substituting the given data into Equation 4.55 yields

$$BP = 0.2(0.15) + 0.5(0.6) + 0.3(0.3)$$
$$= 0.42 percent$$

This result means that the expected sales for the city is 0.42 percent of the total national sale.

Multiattribute Attitude Model   This model is intended to determine the buying behavior of consumers toward a certain product. It is important to note that the model takes the following two items into consideration:

1. Consumer attitude toward each brand of the product.

2. Weight of importance consumers assign to each attribute of the brand.

The following equation is associated with this model [32, 33]:

$$AS_{xy} = \sum_{i=1}^{N}(WT_{iy})AT_{ixy} \qquad \textbf{[4.56]}$$

where

AS_{xy} is the attitude score of consumer y for brand x of the product under consideration.

WT_{iy} is the weight of importance placed on attribute i by consumer y.

AT_{ixy} is the score of the attribute i due to the product brand x because of consumer y's belief.

N is the number of attributes of a given brand.

Assume that a customer of a mechanical product must select between two brands, c and d, and that in the customer's opinion, there are only two important attributes, P and Q. Furthermore, attribute P is two times more important than attribute Q. Table 4.4 presents the scores, out of a maximum of 10, given to attributes P and Q by the customer. Determine the customer's attitude toward both brands. | **Example 4.10**

Table 4.4   **S**cores given by the customer to two product brands

Brand	Attribute P	Attribute Q
c	4	7
d	7	6

Solution

Substituting the given data into Equation 4.56, we get the following for brand c,

$$AS_{c1} = (WT_{P1})AT_{Pc1} + (WT_{Q1})AT_{Qc2}$$
$$= (2)4 + (1)7$$
$$= 15$$

and the following for brand d:

$$AS_{d1} = (WT_{P1})AT_{Pd1} + (WT_{Q1})AT_{Qd1}$$
$$= (2)7 + (1)6$$
$$= 20$$

This result means that the customer is more inclined toward product brand d.

4.6   COMPUTER-AIDED DESIGN (CAD)

CAD is the process of solving design-related problems using computers. This process encompasses the analysis of design data, design-related information storage and retrieval, and the generation and modification of graphic images on a video display [34].

The history of CAD may be traced to the early 1950s when the Servomechanisms Laboratory of the Massachusetts Institute of Technology (MIT) developed an automatically controlled milling machine using the Whirlwind computer [35], ultimately resulting in the evolution of the *automatically programmed tool (APT)* [36]. In 1963, Coons [37] outlined the step from APT to design programs. In the same year, Sutherland [38] of MIT envisaged the engineering designer sitting in front of a console and using interactive graphics facilities called SKETCHPAD. A very early implementation of the idea of having the CAD processing power distributed among local interactive workstations from a central host computer was tried by Bell Telephone Laboratories when it developed the GRAPHIC 1 remote display system [39]. Since the mid-1960s many CAD developments have taken place. However, not until the early 1980s was CAD fully developed in the marketplace. The detailed history of CAD is given in Reference 40.

The use of computers in the design process varies according to the design stage. For example, during the early stages (e.g., needs recognition and problem definition, feasibility study, preliminary design), the applications of the computer are more limited, as these stages represent the more creative aspects of design problem solving (which are generally best performed by the designer).

In contrast, the latter stages (i.e., testing, evaluating, and improving the final design, developing the documentation, communicating the design, defining the manufacturing process, and analyzing the feedback) generally require a large number of repetitive and iterative tasks, which are better suited for computer applications. Some of the design tasks that lend themselves to computer applications are as follows [41]:

1. Repetitive computations or calculations.
2. Complex and time-consuming computations requiring a high level of accuracy.
3. Data manipulation (e.g., the storage and retrieval of information on earlier designs and associated modifications).

It should also be noted that the use of the computer as a design aid is not restricted to the designer; all of the design team members can benefit from its usage.

A CAD system has five basic components: database, program library containing application programs, graphics, data input/output query integrity check, and dialogue. The program library consists of two types of modules. Type I modules are used for primary system functions, including the database, graphics, data input/output, and dialogue. Type II modules contain the algorithms for application areas. The modules for dialogue, input/output data, and graphical information processing form the CAD communications subsystem.

4.6.1 CAD DATABASE

Various types of databases are used by a CAD system. The development of such databases must be carefully considered from different aspects: need, size, application, control, type, etc. Having a central control over the data has several benefits [42]: maintenance of data integrity, easy enforcement of standards, reduction in data redundancy, better consistency, reduction in security restrictions, and sharing of stored data. In the specifications for the CAD database requirements, the following factors should be considered [40]; entities to be stored, relationships between the entities, potential users, simplicity of the resulting algorithms, data-structure operations to be performed, and frequency of the different data management operations.

4.6.2 CAD ECONOMIC JUSTIFICATION

As is the case for any engineering system, the introduction of interactive CAD systems requires economic justification. For this purpose, Chasen and Dow [43] proposed making a distinction between direct and indirect cost benefits. The direct cost benefits include cost reduction and improved productivity ratio. The indirect cost benefits include improved design, better product quality, reduced project development time, improved program interfaces, manpower augmentation, and rapid elimination of impractical approaches. Reference 43 further proposed the following evaluation approach:

1. Estimate the value of the productivity ratio (Rp) using the following equation:

$$Rp = (T_p - T_a)/T_c \qquad \textbf{[4.57]}$$

where

T_c is the time spent at the console, expressed in manhours.

T_a is the time unaffected by CAD, expressed in manhours.

T_p is the time prior to the introduction of CAD, expressed in manhours.

2. Estimate the reduction in cost (C_r) using the following relationship:

$$C_r = IC_b + (T_p - T_a)C_p - T_cC_c - T_cC_p \qquad \textbf{[4.58]}$$

where

IC_b is the estimated cost of the indirect benefits (under the worst-case condition $IC_b = 0$).

C_p is the hourly average personnel cost.

C_c is the hourly console rate cost.

3. Use Equations 4.57 and 4.58, along with the following typical constraints imposed by the market and company policy, in making the decision regarding the introduction of CAD to an organization:

 a. Competition.
 b. Level of allowable investment.
 c. Innovation related pressure.
 d. Past experience.
 e. Company priorities.
 f. Form of the organization.

4.6.3 CAD ADVANTAGES

Over the years, many benefits have been realized from using a CAD system. Some of these are [34, 43]:

1. Improvement in accuracy.
2. Simplification in creating and correcting working drawings.
3. Efficient and ideal solution of computational design-analysis problems.
4. Effective simulation and testing of designs under consideration.
5. Reduction in the design process steps.
6. Improvement in designer's productivity.
7. Easier graphical manipulation of the proposed design.

4.7 PROBLEMS

1. Describe the following terms:
 a. Lagrange multiplier.
 b. Linear programming.
 c. Critical path.
 d. Optimization.
2. Discuss the group brainstorming method.
3. List and discuss the attributes of a creative design engineer.
4. Discuss the steps associated with the creative problem-solving process.
5. Describe at least five important guidelines for improving creative thinking.

6. Discuss at least eight important functions of marketing.

7. Using the brainstorming technique, generate ideas to minimize the cost of lighting a factory.

8. Assume that the reliability of an engineering system is expressed by

$$R_s = (1 - F_c)^k - F_o^k \qquad \qquad [4.59]$$

where

R_s is the system reliability.

F_c is the device closed-mode failure probability.

F_o is the device open-mode failure probability.

K is the number of devices.

Compute the optimum number of devices that will maximize the system reliability for open-mode and closed-mode failure probabilities of 0.15 and 0.20, respectively.

9. List the causes of product failure in the marketplace.

10. Assume that a company manufactures two types, A and B, of an engineering product, which uses three different types of parts X, Y, and Z. The company has 30 parts of type X, 40 parts of type Y, and 25 parts of type Z, respectively. Product type A requires five type Y parts and four type X parts. Product type B requires six type Z parts and two type Y parts. If each product type A makes a profit of $30 and each product type B makes a profit of $20, determine graphically how many units of each product type to manufacture to maximize profit.

11. Assume that an engineering course term report project is broken down into a number of major jobs or activities, as shown in Table 4.5. Develop a CPM network using the data in the table, and determine the critical path of the network, along with the project duration time period.

Table 4.5   **A**n engineering course term report activity breakdown and activity durations

Activity No.	Activity Description	Activity Identification	Immediate Predecessor Activity	Activity Duration (days)
1	Literature Collection	A	-	7
2	Literature review	B	A	4
3	Outline preparation	C	B	1
4	Analysis	D	B	10
5	Report writing	E	C	5
6	Typing	F	E	3
7	Review and revision	G	E	4
8	Final draft preparation	H	G	2

REFERENCES

1. Siddall, J.N. "Mechanical Design." *American Society of Mechanical Engineers (ASME) Transactions: Journal of Mechanical Design* 101, (1979), pp. 674–681.

2. Dieter, G. *Engineering Design.* New York: McGraw-Hill, 1983.

3. Walton, J. *Engineering Design: From Art to Practice.* New York: West Publishing Company, 1991.

4. Meredith, D.D.; K.W. Wong; R.W. Woodhead; and R.H. Wortman *Design and Planning of Engineering Systems.* Englewood Cliffs, NJ: Prentice-Hall, 1985.

5. Dhillon, B.S. *Quality Control, Reliability and Engineering Design.* New York: Marcel Dekker, 1985

6. Stoecker, W.F. *Design of Thermal Systems.* New York: McGraw-Hill, 1980.

7. Lee, S.M.; L.J. Moore; and B.W. Taylor; *Management Science.* Dubuque, IA: Wm. C. Brown Company, 1981.

8. Buffa, E.S. *Modern Production/Operations Management.* New York: Wiley, 1983.

9. Riggs, J.L.; and M.S. Inoue. *Introduction to Operations Research and Management Science: A General Systems Approach.* New York: McGraw-Hill, 1975.

10. Dhillon, B.S. *Engineering Management: Concepts, Procedures and Models.* Lancaster, PA: Technomic Publishing Company, 1987.

11. Malcolm, D.G.; J.H. Roseboom; C.E. Clark; and W. Fazar. "Application of a Technique for Research and Development Program Evaluation." *Operations Research* 7, (1959), pp. 646–669.

12. Gaither, N. "The Adoption of Operations Research Techniques by Manufacturing Organizations." *Decision Sciences* 6, (1975), pp. 794–814.

13. Chase, R.B.; and N.J. Aquilano. *Production and Operations Management: A Life Cycle Approach.* Homewood, IL: Richard D. Irwin, Inc., 1981.

14. Dhillon, B.S. *Reliability Engineering in System Design and Operation.* New York: Van Nostrand Reinhold, 1983.

15. Clark, C.E. "The PERT Model for the Distribution of an Activity Time." *Operations Research* 10, (1962), pp. 405–406

16. Rothenberg, A.; and B. Greenberg. *The Index of Scientific Writings on Creativity.* Hamden, CT: The Shoe String Press, 1976.

17. Ullman, D.G. *The Mechanical Design Process.* New York: McGraw-Hill, 1992.

18. Osborn, A.F. *Applied Imagination.* New York: Charles Scribner and Sons, 1963.

19. Alger, J.R.M.; and C.V. Hays. *Creative Synthesis in Design.* Englewood Cliffs, NJ: Prentice-Hall, 1964.

20. Van Frange, E. *Professional Creativity.* Englewood Cliffs, NJ: Prentice-Hall, 1959.

21. Bronikowski, R.J. "Creativity Steps." *Chemical Engineer,* July 31, 1978, pp. 103–108.

22. Ray, M. S. *Elements of Engineering Design.* Englewood Cliffs, NJ: Prentice-Hall, 1985.

23. Shannon, R.E. *Engineering Management.* New York: Wiley, 1980.

24. Dhillon, B.S. *Engineering Management.* Lancaster, PA: Technomic Publishing Company, 1987.

25. Beakley, G.C.; and E.G. Chilton. *Introduction to Engineering Design and Graphics.* New York: Macmillan, 1973.

26. Gordon, W.J. *Synectics.* New York: Harper and Brothers, 1961.

27. Haefele, J.W. *Creativity and Innovation.* New York: Reinhold, 1962.

28. Lanigan, M. *Engineers in Business.* Wokingham, UK: Addison-Wesley, 1992.

29. Hise, R.T.; P.L. Gillett; and J.K. Ryans. *Basic Marketing: Concepts.* Cambridge, MA: Winthrop Publishers, Inc., 1979.

30. Diamond, J.; and G. Pintel. *Principles of Marketing.* Englewoods Cliffs, NJ: Prentice-Hall, 1972.

31. Turck, F.B. "The Forty-Seven Functions of Marketing." In *Management Guide for Engineers and Technical Administrators,* ed. N.P. Chironis, New York: McGraw-Hill, 1969, pp. 300–303.

32. Kotler, P. *Marketing Management: Analysis, Planning and Control.* Englewood Cliffs, NJ: Prentice-Hall, 1980.

33. Wilkie, W.L.; and E.A. Pessemier. "Issues in Marketing's Use of Multi-Attribute Attitude Models." *Journal of Marketing Research,* 1973, pp. 428–441.

34. Earle, J.H. *Engineering Design Graphics.* Reading, MA: Addison-Wesley, 1990.

35. Pease, W. "An Automatic Machine Tool." *Scientific American* 187, (1952), pp. 101–115.

36. Brown, S.; C. Drayton; and B. Mittman. "A Description of the APT-Language." *CACM* 6, (1963), pp. 649–658.

37. Coons, S. "An Outline of the Requirements for a Computer-Aided Design System." *AFIPS (SJCC)* 23, (1963) pp. 299–304.

38. Sutherland, I. "SKETCHPAD: A Man-Machine Graphical Communication System." *AFIPS (SJCC)* 23 (1963) pp. 329–346.

39. Ninke, W. "GRAPHIC 1: A Remote Graphical Display Console System." *AFIPS (SJCC)* 22, (1965) pp. 839–846.

40. Encarnacao, J.; and E.G. Schlechtendahl. *Computer Aided Design.* Berlin: Springer-Verlag, 1983.

41. Ray, M.S. *Elements of Engineering Design: An Integrated Approach.* Englewood Cliffs, NJ: Prentice-Hall, 1985.

42. Date, C.J. *An Introduction to Database Systems.* Reading, MA: Addison-Wesley, 1976.

43. Chasen, S.H.; and J. Dow. *The Guide for the Evaluation and Implementation of CAD Systems.* Atlanta, GA: CAD Decisions 1979.

5

PROBABILITY AND STATISTICS

5.1 INTRODUCTION

Usually, during the design phase of an engineering product, the involved professionals must deal with various types of data, which must be carefully analyzed. A knowledge of probability and statistics therefore plays an important role in this analysis. Questions which could be effectively answered using probability and statistics are as follows [1]:

1. What statistical distribution effectively represents the experimental data?

2. Is equipment X better than equipment Y?

3. How can the available table-form data be effectively used in the design work?

4. What is the real meaning of the available table-form materials property data?

5. Do the samples taken belong to the same batch?

6. After how many hours of operation can we expect the product to fail?

The history of probability goes back to the sixteenth century when Girolamo Cardano, (1501–1576), an Italian, wrote a gambler's guidebook containing some interesting questions on probability [2]. However, the real advance in the development of probability occurred in 1654, in France, when Chevalier de Méré, an experienced gambler, consulted Blaise Pascal (1623–1662), a mathematician, on how to divide the winnings in a game of chance. Pascal became quite interested in the problem and transmitted it to Pierre Fermat (1601–1665). The problem was solved correctly and independently by these two men. Other significant contributions to the field of probability were made by Karl Gauss (1777–1855) and Pierre Laplace (1749–1827). Since then, many other researchers and authors have made contributions to the field, and today probability theory is applied to many diverse areas, of which engineering design is one.

5.2 MEAN, MEDIAN, MODE, STANDARD DEVIATION AND VARIANCE

5.2.1 MEAN

The *mean* is the sum of all the values of a variable in a sample of that variable divided by the size of the sample. This is expressed mathematically as follows:

$$\overline{T} = \frac{\sum\limits_{i=1}^{m} T_i}{m}$$ [5.1]

where

$\overline{T}$ which is read "T bar," is the mean of the variable values in the sample.

T_i is the value of the ith variable in the sample.

m is the sample size.

Example 5.1 | **A**ssume that a random sample of various times-to-failure of an electric fan was taken over a given period of time. The values of the time-to-failure were: 2000, 2500, 3000, 1500, 1800, 2700, and 1700 hours. Calculate the mean value of the time-to-failure.

Solution

Substituting the given data into Equation 5.1 yields

$$\overline{T} = \frac{(2000) + (2500) + (3000) + (1500) + (1800) + (2700) + (1700)}{7}$$

$$= 2171.5 \; hours$$

Thus, the average or mean value of the fan times-to-failure is 2171.5 hours.

5.2.2   MEDIAN

The *median* is a value taken for a set of observed values arranged in order of magnitude. The median is given by either the middle value or the average of the two middle values, whichever is applicable [3,4].

Example 5.2 | **A** sample of the prices of a mechanical pump manufactured by various manufacturers is given in Table 5.1. Determine the value of the median.

Solution

If the cost figures in Table 5.1 are arranged in order of magnitude, they are: 9000, 10,000, 11,000, 12,000, 14,000, 15,000, and 16,000. The value of the median is therefore $12,000.

Example 5.3 | **A**ssume that in Example 5.2 an additional observation was obtained, yielding manufacturer H and $13,000. Determine the new value of the median.

Table 5.1 Pump costs

Manufacturer	Cost (dollars)
A	10,000
B	15,000
C	11,000
D	9,000
E	16,000
F	14,000
G	12,000

Solution

The pump prices in order of magnitude become: 9,000, 10,000, 11,000, 12,000, 13,000, 14,000, 15,000, and 16,000. In this case, the value of the median M is

$$M = \frac{(12,000) + (13,000)}{2} = \$12,500$$

5.2.3 MODE

The *mode* of a set of observations in a sample is the most frequently occurring observation. The following concepts are associated with the mode:

1. There may not be a mode for the set of observations (or, there may be more than one mode).
2. For a symmetrical distribution, the values of the mode, median, and mean are the same.
3. For positively skewed distributions, the value of the mode is less than that of the median.
4. For negatively skewed distributions, the value of the mode is greater than that of the median.

A set of data having only one mode is called "unimodal"; if there are two modes, the data set is called "bimodal."

Determine if the data set given in Example 5.1 has a mode.   **Example 5.4**

Solution

The data set in that example has no mode.

5.2.4 STANDARD DEVIATION

The *standard deviation* is a useful measure of variability. For a random sample, the standard deviation is calculated using the following relationship [3,4]:

$$SD = \left(\frac{\sum\limits_{i=1}^{m} (T_i - \overline{T})^2}{m - 1} \right)^{1/2}$$ **[5.2]**

where

  SD is the standard deviation of the random sample.

To obtain the standard deviation of a population (set of variables), replace SD, $\overline{T}$, and m with σ, μ, and M, respectively, in 5.2. The symbols σ, μ, and M refer to the population standard deviation, mean, and size, respectively.

The definition of a *normal distribution* is that 68.27 percent of the area under the curve is between $(\mu + \sigma)$ and $(\mu - \sigma)$, 95.45 percent is between $(\mu + 2\sigma)$ and $(\mu - 2\sigma)$, and 99.73 percent is between $(\mu + 3\sigma)$ and $(\mu - 3\sigma)$.

Example 5.5

The sampled tolerances [in millimeters (mm)] of certain types of bearings are given in Table 5.2. Calculate the standard deviation of the sample.

Solution

Substituting the Table 5.2 data into Equation 5.1 yields

$$\overline{T} = \frac{0.34}{8} = 0.0425 \ mm$$

Table 5.2   **B**earing tolerances

Observation No.	Tolerance (mm)
1	0.06
2	0.05
3	0.05
4	0.01
5	0.07
6	0.03
7	0.02
8	0.05

Using this result and the data in Table 5.2 in Equation 5.2, we get

$$SD = \left[\frac{\begin{array}{c}(0.06 - 0.0425)^2 + (0.05 - 0.0425)^2 + \\ (0.05 - 0.0425)^2 + \cdots + (0.05 - 0.0425)^2\end{array}}{(8 - 1)} \right]^{1/2} = 0.0205mm$$

Thus, the standard deviation of the sample is 0.0205 mm.

5.2.5 VARIANCE

The *variance* is a measure of the dispersion of the values about the mean. The variance is defined as the square of the standard deviation. From Equation 5.2, the variance of the random sample is

$$(SD)^2 = \frac{\sum_{i=1}^{m}(T_i - \overline{T})}{(m - 1)} \qquad \textbf{[5.3]}$$

Similarly, the population variance is

$$\sigma^2 = \left(\frac{\sum_{i=1}^{M}(T_i - \mu)}{M - 1} \right) \qquad \textbf{[5.4]}$$

Calculate the value of the variance for the Example 5.5 data given in Table 5.2.   **Example 5.6**
Solution
Substituting the Table 5.2 data into Equation 5.3 yields

$$(SD)^2 = (0.0205)^2 = 0.00042 \; mm^2$$

5.3 PROBABILITY

Probability may be defined as the study of random experiments. The occurrence probability of event x is expressed as

$$p(x) = \lim_{n \to \infty} \left(\frac{N}{n} \right) \qquad \textbf{[5.5]}$$

where

$p(x)$ is the probability of the occurrence of event x.

N is the number of times that event x actually occurs.

n is the number of times an experiment is repeated.

In practice, it is usually impossible to conduct a very large number of repeated experiments; therefore, the value of $p(x)$ is simply approximated by (N/n).

Some of the properties, or axioms, of probability are:

1. For every event x, the probability of x occurring is in the range $0 \leq p(x) \leq 1$.

2. The probability of the sample space S is

$$p(S) = 1$$

3. The probability of the union of m mutually exclusive events, $x_1, x_2, x_3, \ldots, x_m$, is given by

$$p(x_1 + x_2 + \ldots + x_m) = p(x_1) + p(x_2) + \ldots + p(x_m) \qquad \textbf{[5.6]}$$

4. Where the symbol $(+)$ denotes a union of events, and $p(x_i)$ is the probability of occurrence of event x_i, for $i = 1, 2, 3 \ldots, m$. For non-mutually exclusive events, such as x_1 and x_2, the relationship becomes

$$p(x_1 + x_2) = p(x_1) + p(x_2) - p(x_1)p(x_2). \qquad \textbf{[5.7]}$$

5. The probability of intersections of m independent events is

$$p(x_1 x_2 x_3 \ldots x_m) = p(x_1)p(x_2)p(x_3) \ldots p(x_m) \qquad \textbf{[5.8]}$$

5.4 DISCRETE PROBABILITY DISTRIBUTIONS

Two important *discrete-random-variable distributions* are the *binomial distribution* and *Poisson distribution*. These are discussed in the following sections.

5.4.1 BINOMIAL DISTRIBUTION

Also known as the *Bernoulli distribution*, the binomial distribution is used in those situations in which each trial has two possible outcomes e.g., success or failure, good or bad. In addition, each trial probability must remain constant. The binomial distribution is often used in quality control and reliability applications.

The *binomial probability function $f(y)$* is defined as

$$f(y) = \frac{m!}{y!(m-y)!} p^y q^{m-y} \qquad for \ y = 0, 1, 2, 3, \ldots, m \qquad \textbf{[5.9]}$$

where

y is the number of occurrences in m trials.

q is the occurrence probability of each trial.

p is the nonoccurrence probability of each trial.

The *cumulative probability distribution function $F(y)$* is given by

$$F(y) = \sum_{i=0}^{y} \left[\frac{m!}{y!(m-y)!} \right] p^i q^{m-i} \qquad \textbf{[5.10]}$$

where

$F(y)$ is the probability of y or less occurrences in m trials.

Note that each trial's probability of occurrence (q) plus its probability of non-occurrence (p), is always equal to unity.

The expressions for the variance, mean, and standard deviation, respectively, of the binomial distribution are as follows:

$$\sigma^2 = pqm \qquad \textbf{[5.11]}$$

$$\mu = pm \qquad \textbf{[5.12]}$$

$$\sigma = [pqm]^{1/2} \qquad \textbf{[5.13]}$$

The symbols σ^2, μ, and σ represent the binomial distribution variance, mean, and standard deviation, respectively.

Assume that for a resistor manufacturer, 15 percent of the items produced are out of tolerance. Calculate the probability of having four rejected resistors out of a sample of eight resistors. | **Example 5.7**

Solution

Substituting the given data into Equation 5.9 yields

$$f(4) = \frac{8!}{4!(8-4)!}(0.15)^4(0.85)^{8-4}$$

$$= \frac{40320}{576}(0.15)^4(0.85)^4$$

$$= 0.019$$

The result means that the probability of having four rejected resistors out of a sample of eight is approximately 2 percent.

5.4.2 POISSON DISTRIBUTION

This distribution is closely related to the binomial distribution in that, for an infinite number of trials m, the binomial distribution becomes Poisson distribution, because p approaches zero [5]. Poisson distribution is useful in designing systems in which the number of units to be handled per unit of time is important, as well as the process flow (flow per unit of length), such as in sheet steel, paper, and cloth manufacture. The distribution was discovered by Poisson in the early part of the 19th century.

The probability density function is expressed as

$$f(k) = \frac{(\lambda t)^k e^{-\lambda t}}{k!}$$ **[5.14]**

where

t is the time.

λ is the occurrence rate per unit of time.

k is the number of occurrences per unit of time.

The cumulative distribution function is

$$F = \sum_{i=0}^{k}(\lambda t)^i \, e^{-\lambda t}/i!$$ **[5.15]**

Expressions for the mean, standard deviation, and variance, respectively, of Poisson distribution, are as follows:

$$\mu = \lambda$$ **[5.16]**

$$\sigma = \sqrt{\lambda}$$ **[5.17]**

$$\sigma^2 = \lambda$$ **[5.18]**

Example 5.8

Assume that the average rate of manufacturing television sets at a production line is 10 sets per hour. Calculate the probability of manufacturing exactly 10 television sets in the next hour.

Solution

In this case, we have $\lambda = 10$ television sets/hour, $t = 1$ hour, and $k = 10$. Substituting these data into Equation 5.14 yields

$$f(10) = \frac{[(10)(1)]^{10}e^{-(10)(1)}}{(10)!}$$

$$= 0.125$$

This result means that the probability of manufacturing exactly 10 television sets in the next hour is approximately 12.5 percent.

5.5 CONTINUOUS PROBABILITY DISTRIBUTIONS

There are many *continuous-random-variable probability distributions*, such as the *normal distribution, exponential distribution, Rayleigh distribution,* and *Weibull distribution.* The probability density function for continuous distributions is defined as

$$f(x) = \frac{dF(x)}{dx} \qquad \textbf{[5.19]}$$

where

$$F(x) = \int_{-\infty}^{x} f(t)dt \qquad \textbf{[5.20]}$$

$$F(\infty) = 1 \qquad \textbf{[5.21]}$$

The term $F(x)$ is known as the *cumulative distribution function* of the continuous random variable x. The distribution mean μ and variance σ^2 are defined as follows [6]

$$\mu = \int_{-\infty}^{\infty} xf(x)dx \qquad \textbf{[5.22]}$$

$$\sigma^2 = \int_{-\infty}^{\infty} (t - \mu)^2 f(x)dx \qquad \textbf{[5.23]}$$

5.5.1 NORMAL DISTRIBUTION

The normal distribution, also known as the *Gaussian distribution,* is widely used. It concerns the distribution of two parameters and is defined as

$$f(x) = \frac{1}{\sigma\sqrt{2\pi}}\exp\left[-\frac{1}{2}\left(\frac{x-\mu}{\sigma}\right)^2\right] \qquad -\infty < x < \infty \qquad \textbf{[5.24]}$$

where

$f(x)$ is the probability density function of the continuous random variable x.

μ is the mean distribution parameter.

σ is the standard deviation distribution parameter.

The cumulative distribution function from Equations 5.20 and 5.22 is given by

$$F(t) = \frac{1}{\sigma\sqrt{2\pi}}\int_{-\infty}^{t} \exp\left[-\frac{1}{2}\left(\frac{x-\mu}{\sigma}\right)^2\right]dx \qquad \textbf{[5.25]}$$

The numerical values of Equation 5.23 can be obtained from standard normal tables. For $\mu = 0$ and $\sigma = 1$, the Gaussian distribution is known as the *standard normal distribution*. Thus, from Equation 5.24, we have

$$f(x) = \frac{1}{\sigma\sqrt{2\pi}} \exp\left[-\frac{1}{2}x^2\right] \qquad \textbf{[5.26]}$$

Table 5.3 presents selected values of the standard normal cumulative distribution function.

Table 5.3   **S**elected values of the standard normal cumulative distribution function $F(t)$

t	$F(t)$	t	$F(t)$
−3.5	0.00017	0.5	0.692
−3	0.001	1	0.841
−2.5	0.006	1.5	0.933
−2	0.023	2	0.977
−1.5	0.067	2.5	0.994
−1	0.159	3	0.999
−0.5	0.309	3.5	0.99977
0	0.5		

Example 5.9

A number of shafts with a diameter of 100 millimeters were manufactured by a company. The distribution of the shaft diameter is normal with $\mu = 100.05$ mm and $\sigma = 1$ mm. Calculate the probability that shafts selected at random will have a diameter greater than 101 mm.

Solution

The value of t is given by

$$t = \frac{101 - 100.5}{1} = 0.5$$

Using Table 5.3, we obtain the corresponding probability value 0.692. Subtracting this result from unity yields 0.308, which means that there is 30.8 percent chance of selecting shafts with diameters greater than 101 mm.

5.5.2   EXPONENTIAL DISTRIBUTION

The exponential distribution function is widely used in conducting reliability analyses. The probability density function is defined as

$$f(x) = \lambda e^{-\lambda x} \qquad \lambda > 0, x \geq 0 \qquad \textbf{[5.27]}$$

where

x is the random variable. (In reliability work, x denotes time.)

λ is the distribution parameter. (In reliability work, this represents the failure rate of an item.)

The cumulative distribution function, from Equations 5.20 and 5.25, is

$$F(x) = 1 - e^{-\lambda x} \qquad \textbf{[5.28]}$$

Expressions for the distribution mean μ and variance σ^2, using Equations 5.22, 5.23, and 5.27, are

$$\mu = \frac{1}{\lambda} \qquad \textbf{[5.29]}$$

$$\sigma^2 = \frac{1}{\lambda^2} \qquad \textbf{[5.30]}$$

A newly designed air compressor was tested over a period of time. Its times-to-failure follow the exponential distribution, with a constant failure rate of 0.0004 failures/hour. Calculate the failure probability of the compressor for a 200-hour mission. | **Example 5.10**

Solution

Substituting the given data into Equation 5.28 yields,

$$
\begin{aligned}
F(200) &= 1 - e^{-(0.0004)(200)} \\
&= 0.077
\end{aligned}
$$

This result means there is a 7.7 percent chance that the air compressor will malfunction during the 200-hour operation.

5.5.3 RAYLEIGH DISTRIBUTION

The Rayleigh distribution is often used in the theory of sound and reliability engineering. In reliability work, this distribution is used to represent the failure behavior of items whose failure rate increases linearly [5].

The distribution probability density function is defined as

$$f(x) = \frac{2x}{\beta^2} \exp\left[-\left(\frac{x}{\beta}\right)^2\right] \qquad \beta > 0,\ x \geq 0 \qquad \textbf{[5.31]}$$

where

β is the scale parameter.

The cumulative distribution function, from Equations 5.20 and 5.31, results in

$$F(x) = 1 - \exp\left[-\left(\frac{x}{\beta}\right)^2\right] \qquad \textbf{[5.32]}$$

Substituting Equation 5.31 into Equations 5.22 and 5.23 leads to the following expressions for the distribution mean μ and variance σ^2.

$$\mu = \frac{1}{\beta} \Gamma \left(\frac{3}{2} \right) \qquad \textbf{[5.33]}$$

where

$\Gamma(\cdot)$ is known as the gamma function, which is defined as

$$\Gamma(n) = \int_0^\infty x^{n-1} e^{-x} dx \qquad n > 0 \qquad \textbf{[5.34]}$$

and

$$\sigma^2 = \frac{1}{\beta^2} \left\{ \Gamma(2) - \left[\Gamma \left(\frac{3}{2} \right) \right]^2 \right\} \qquad \textbf{[5.35]}$$

Example 5.11 | **A** newly designed electric motor was tested for reliability over a period of time. Its times-to-failure can be described by the Rayleigh distribution with a scale parameter value of 400. Calculate the failure probability of the motor for a 100-hour mission.

Solution

Substituting the given data into Equation 5.32 yields

$$F(100) = 1 - \exp \left[-\left(\frac{100}{400} \right)^2 \right] = 0.0606$$

This result means that there is approximately a 6 percent chance that the motor will fail during the 100-hour mission.

5.5.4 WEIBULL DISTRIBUTION

The Weibull distribution can be used to represent many different physical phenomena. It is named after its originator, W. Weibull [6]. The Weibull distribution is often used in reliability studies. The probability density function is defined as

$$f(x) = \frac{b}{\beta} \left(\frac{x}{\beta} \right)^{b-1} \exp \left[-\left(\frac{x}{\beta} \right)^b \right] \qquad \beta, b > 0, \ t \geq 0 \qquad \textbf{[5.36]}$$

where

β is the scale parameter.
b is the shape parameter.

For $b = 1$ and $b = 2$, the Weibull distribution becomes the exponential distribution and the Rayleigh distribution, respectively.

The cumulative distribution function, from Equations 5.20 and 5.36, is

$$f(x) = 1 - \exp\left[-\left(\frac{x}{\beta}\right)^b\right] \qquad \textbf{[5.37]}$$

Using Equations 5.22, 5.23, and 5.36, we get the following expressions for the distribution mean μ and variance σ^2:

$$\mu = \frac{1}{\beta}\Gamma\left(\frac{1}{b} + 1\right) \qquad \textbf{[5.38]}$$

$$\sigma^2 = \frac{1}{\beta^2}\left\{\Gamma\left(1 + \frac{2}{b}\right) - \left[\Gamma\left(1 + \frac{1}{b}\right)\right]^2\right\} \qquad \textbf{[5.39]}$$

The times-to-failure of an aircraft engine, as observed over a period of time, are Weibull distributed, with a shape parameter value of 1 and a scale parameter of 500 hours. Calculate the failure probability of the engine for a 15-hour mission.

| **Example 5.12**

Solution

Substituting the given data into Equation 5.37 yields

$$F(15) = 1 - \exp\left[-\left(\frac{15}{500}\right)\right]$$
$$\simeq 0.03$$

This result means that there is only a 3 percent chance that the engine will cease to operate during the 15-hour mission.

5.6 STATISTICAL TESTS

In the published literature, various types of statistical tests have been developed. In this section, we discuss two of these tests: the Bartlett Test, and the z-test.

5.6.1 BARTLETT TEST

In design work, the given data often follow an exponential distribution, especially when reliability studies are performed. The *Bartlett test* is used to determine whether or not the data follow an exponential distribution. The Bartlett test statistic is defined [7–9] as

$$S_m = 12m^2\left[\ln P - \frac{Q}{m}\right]/(6m + m + 1) \qquad \textbf{[5.40]}$$

where

$$P \equiv \frac{1}{m} \sum_{j=1}^{m} x_j \qquad \qquad \textbf{[5.41]}$$

and

$$Q = \sum_{j=1}^{m} \ln x_j \qquad \qquad \textbf{[5.42]}$$

where

x_j is the value (i.e., time-to-failure) of the jth element in a sample, and m is the total number of observations in the sample.

The effectiveness of this test depends on the sample size; for the test to be effective, the recommended sample size is 20 or more. The test is based on the reasoning that, if the observation values (e.g., failure times) are exponentially distributed, then S_m is distributed as chi-square with $(m-1)$ degrees of freedom; thus, a two-tailed chi-square criterion is used [8]. The selected critical approximate values of the chi-square distribution are given in Table 5.4. More accurate values of the chi-square distribution may be found in Reference [10].

Example 5.13 | **A**ssume that a computer system was sampled 24 times to determine the times-to-failure and the results are given in Table 5.5. Use the Bartlett test to determine if the sampled data demonstrate an exponential distribution.

Solution

Substituting the Table 5.5 data into Equation 5.41 yields

$$P = \frac{1}{24}[6 + 20 + 9 + 14 + 25 + 30 + 34 + 41 + 67 + 60 + 46 + 55 + 90 + 95$$

$$+ 108 + 110 + 150 + 145 + 150 + 255 + 260 + 190 + 195 + 180]$$

$$= 97.29$$

Similarly, from Equation 5.42, we get

$$Q = [1.792 + 2.996 + 2.197 + 2.639 + 3.219 + 3.401 + 3.526 + 3.71$$

$$+ 4.205 + 4.094 + 3.829 + 4.007 + 4.5 + 4.554 + 4.682 + 4.701$$

$$+ 5.011 + 4.977 + 5.561 + 5.247 + 5.273 + 5.193]$$

$$= 99.87$$

Inserting these results into Equation 5.40 yields

$$S_m = 12(24)^2 \left[\ln 97.29 - \frac{99.87}{24} \right] / \{(6)(24) + 24 + 1\}$$

$$= 17.03$$

Table 5.4   Values of the chi-square distribution

Degrees of Freedom	Probability (α)					
	0.99	0.975	0.95	0.05	0.02	0.01
1	0.0002	0.001	0.004	3.8	5.0	6.6
2	0.02	0.05	0.1	6	7.4	9.2
3	0.12	0.22	0.35	7.8	9.4	11.3
4	0.30	0.48	0.7	9.5	11.1	13.3
5	0.55	0.83	1.2	11.1	12.8	15.1
6	0.87	1.2	1.6	12.6	14.5	16.8
7	1.2	1.7	2.2	14.1	16.0	18.5
8	1.7	2.2	2.7	15.5	17.5	20.1
9	2.1	2.7	3.3	16.9	19.0	21.7
10	2.6	3.3	3.9	18.3	20.5	23.2
11	3.1	3.8	4.6	19.7	21.9	24.7
12	3.6	4.4	5.2	21	23.3	26.2
13	4.1	5.0	5.9	22.4	24.7	27.7
14	4.7	5.6	6.6	23.7	26.1	29.1
15	5.2	6.3	7.3	24.0	27.5	30.6
16	5.8	6.9	8	26.3	28.9	32.0
17	6.4	7.6	8.7	27.6	30.2	33.4
18	7.0	8.2	9.4	28.9	31.5	34.8
19	7.6	8.9	10.1	30.1	32.9	36.2
20	8.3	9.6	10.9	31.4	34.2	37.6
21	8.9	10.3	11.6	32.7	35.5	38.9
22	9.5	11.0	12.3	33.9	36.8	40.3
23	10.2	11.7	13.1	35.2	38.1	41.6
24	10.9	12.4	13.9	36.4	39.4	43.0
25	11.5	13.1	14.6	37.7	40.7	44.3

Table 5.5   Computer system times-to-failure observations (in days)

6	34	90	150
20	41	95	255
9	67	108	260
14	60	110	190
25	46	150	195
30	55	145	180

Using Table 5.4 for a two-tailed test with a 90 percent confidence level, we obtain the following critical values:

$$\chi^2\left[\frac{\beta}{2}, (m-1)\right] = \chi^2\left[\frac{0.1}{2}, (24-1)\right] = 35.2$$

where

$\beta = 1-(\text{confidence level}) = 1 - 0.90 = 0.1$, and

$\chi^2(\cdot)$ is the chi-square distribution.

Also,

$$\chi^2\left[\left(1-\frac{\beta}{2}\right), (m-1)\right] = \chi^2\left[\left(1-\frac{0.1}{2}\right), (24-1)\right] = 13.1$$

Since the value of $S_m = 17.03$ falls between 35.2 and 13.1, it is safe to assume an exponential distribution.

5.6.2   Z-TEST

The z-test is useful for determining whether the sampled data belong to a known or assumed universal group [1]. An example might be a shipment of parts from a vendor and the subsequent inspection of those parts. The z-test assumes that the mean and standard deviation for the universal group are known.

To apply this test, we calculate the value of z using the following equation:

$$z = |\bar{x} - \mu_p| / \left(\sigma_p/\sqrt{m}\right) \qquad\qquad \textbf{[5.43]}$$

where

$\bar{x}$ is the mean of the sampled data values.

μ_p is the mean of the universe or population.

σ_p is the standard deviation of the universe or population.

m is the total number of sampled data values.

In order to use Table 5.3, replace t with z; for reliable results, it is recommended that the value for m be $m \geq 30$. Also, for a small sample size, use the following equation instead of Equation 5.43:

$$z = |\bar{x} - \mu_p| \, (M-1)^{1/2} / \left(\sigma_p/\sqrt{m}\right) (M-m)^{1/2} \qquad\qquad \textbf{[5.44]}$$

where

M is the size of the universe or population.

Table 5.3 presents the selected probability of z being valid, thus indicating whether the sampled data belong to the universe or population. For more information on the z-test, see References 11 and 12.

Assume that the data in Table 5.5 were taken from a population size of 700 with a known mean μ_p and standard devation σ_p of 150 and 20, respectively. Determine whether the sampled data belong to the population.

| **Example 5.14**

Solution

Substituting the given data into Equation 5.44 yields

$$z = |\ 97.29 - 150\ |\ (700 - 1)^{1/2}/(20/\sqrt{24})(700 - 24)^{1/2}$$
$$= 13.13$$

Note that in Table 5.3, the highest value of z or t (i.e., 3.5) is associated with the 0.99977 probability. Since $13.13 > 3.5$, this means that the chance is less than 23 out of 100,000 that the sampled data belong to the universe or population.

5.7   CONFIDENCE LIMITS

In some situations, a population parameter is given as an estimate between two numbers. The two numbers are called the upper and lower confidence limits. The estimate is known as the parameter interval estimate. One example of such a population parameter is the mean value.

In design reliability studies, this type of information can be very useful. This section discusses establishing the confidence limits on the exponential mean life [13]. The discussion presents confidence limit formulations for two different test approaches.

5.7.1   TEST APPROACH TYPE I

In this case, the identical parts or items are tested until a given number of failures occurs. Thus, the lower or one-sided confidence limit is expressed [13] by

$$\left[\frac{2T}{\chi^2(\alpha, 2f)}, \infty \right] \qquad \textbf{[5.45]}$$

The two-sided (i.e., upper and lower) confidence limits are

$$\left[\frac{2T}{\chi^2\left(\frac{\alpha}{2}, 2f\right)}, \frac{2T}{\chi^2\left(1 - \frac{\alpha}{2}, 2f\right)} \right] \qquad \textbf{[5.46]}$$

where

$\chi^2(\cdot)$ denotes the chi-square distribution.

α is the probability {i.e., $\alpha = 1-$(confidence level)} that the true mean time to failure will not fall within the confidence interval.

f is the total number of failures.

The values of T for replacement tests (i.e., failed items repaired or replaced) and nonreplacement tests (i.e., failed items not repaired or replaced), respectively, are defined as

$$T = nt_c \qquad\qquad \textbf{[5.47]}$$

$$T = \sum_{i=1}^{f} t_i + t_c(n - f) \qquad\qquad \textbf{[5.48]}$$

where

n is the total number of items selected for testing.

t_c is the time at the termination of life test.

t_i is the time of failure i.

Example 5.15 | **A**ssume that 30 identical resistors were tested until the tenth failure and each failed resistor was replaced as it failed. The last resistor failed at 400 hours. Calculate the exponential minimum mean time to failure of the resistors (i.e., lower or one-sided confidence limit) with a 95 percent confidence level.

Solution

Substituting the given data into Equation 5.47 results in

$$T = (30)(400) = 12,000 \; hours$$

The acceptable risk of error, or probability, that the true mean time to failure of the resistors will not fall within the confidence interval is

$$\alpha = 1- \; (confidence \; level) = 1 - 0.95 = 0.05$$

Substituting these results and the given data into Equation 5.45 and then using Table 5.4, we get

$$\left[\frac{2(12,000)}{\chi^2\{0.05 - 2(10)\}}, \infty \right] = \left[\frac{24,000}{31.4}, \infty \right]$$
$$= [764.33, \infty]$$

This result means that at the 95 percent confidence level, the minimum value of the resistor mean time to failure is 764.33 hours.

5.7.2 TEST APPROACH TYPE II

This test approach calls for the termination of testing at a specified number of test hours. The formulas for the one-sided (i.e., lower limit) and two-sided (i.e., upper and lower limits) confidence limits, respectively, are as follows [13]:

$$\left[\frac{2T}{\chi^2(\alpha, 2f + 2)}, \infty \right] \qquad \text{[5.49]}$$

$$\left[\frac{2T}{\chi^2\left(\frac{\alpha}{2}, 2f + 2\right)}, \frac{2T}{\chi^2\left(1 - \frac{\alpha}{2}, 2f\right)} \right] \qquad \text{[5.50]}$$

Note that the symbols used in Equation 5.49 and 5.50 are the same as for Test Approach I.

Example 5.16

Assume that 40 identical bearings were selected for testing at time zero hours and the test was terminated at 200 hours. Also assume that the failed bearings were not replaced, and that during the test, 5 of the 40 bearings failed at 50, 150, 90, 180, and 100 hours. With a 95 percent confidence level, determine the exponential minimum mean time to failure for the bearings.

Solution

Substituting the given data into Equation 5.48 leads to

$$T = 50 + 150 + 90 + 180 + 100 + (40 - 6)(200)$$
$$= 7,370 \ hours$$

The acceptable risk of error, or probability, that the true mean time to failure will not fall within the stated confidence limits is

$$\alpha = 1 - \ (confidence \ level) = 1 - 0.95 = 0.05$$

Applying these results and the given data to Equation 5.49 and then using Table 5.4, we get

$$\left[\frac{2(7370)}{\chi^2(0.05, 2(5) + 2)}, \infty \right] = \left[\frac{14,740}{\chi^2(0.05, 12)}, \infty \right]$$

$$= \left[\frac{14,740}{21}, \infty \right] = [701.91, \infty]$$

This result means that the bearing minimum mean time to failure, at a 95 percent confidence level, is 701.91 hours.

5.8 CURVE FITTING AND LINEAR REGRESSION ANALYSIS

Past experience indicates that a relationship often exists between variables associated with a given set of data. The usual practice is to express such a relationship mathematically, that is, to determine an equation relating the variables in question. A statistical analysis technique known as *regression analysis* is often used to provide the best curve fit (i.e., linear or nonlinear) to the given data. This section discusses the straight-line (i.e., linear) fit to the data. To determine the best straight-line relationship that will approximate the value of the dependent variable, such as y, for any value of the independent variable, x, the *least-squares method* is used. Basically, this method fits the data points with a least-squares line such that the total of the squares of the vertical distances from the points to the resulting straight line is a minimum.

Let us assume that the relationship between variables y and x is as follows [4, 14]:

$$y = \theta x + k \qquad [5.51]$$

where

θ and k are constants, for the slope of the line and the y intercept, respectively.

From Reference 14 the expressions for estimating the values of θ and k are

$$\theta = \frac{m \sum\limits_{i=1}^{m} x_i y_i - \left(\sum\limits_{i=1}^{m} x_i\right)\left(\sum\limits_{i=1}^{m} y_i\right)}{A} \qquad [5.52]$$

where

$$A \equiv m \sum_{i=1}^{m} x_i^2 - \left(\sum_{i=1}^{m} x_i\right)^2$$

and

$$k = \frac{\left(\sum\limits_{i=1}^{m} y_i\right)\left(\sum\limits_{i=1}^{m} x_i^2\right) - \left(\sum\limits_{i=1}^{m} x_i\right)\left(\sum\limits_{i=1}^{m} x_i y_i\right)}{A} \qquad [5.53]$$

The symbol m in Equations 5.52 and 5.53 denotes the number of data points.

The correlation between variables x and y is defined as [4]

$$\alpha = \left[\frac{SX}{SY}\right](\theta) \qquad [5.54]$$

Where SX and SY are the standard deviations of variables x and y, respectively.

The values of α vary between -1 and $+1$. Interpretations of the α values are as follows:

•   The zero value indicates that the variables x and y are not related.

•   The values close to $+1$ or -1 mean that all data values tend to coincide with the established regression straight line.

•   The negative value means that the variables x and y are negatively related. More specifically, it means that the slope of the established regression straight line is negative.

Example 5.17 | **T**able 5.6 presents x and y data associated with a certain mechanical design. Fit a least-squares line to the given data.

Table 5.6 **D**ata for variables x and y

x	y
10	10
15	14
18	18
20	19
23	24
25	26
27	28
30	34
33	38
35	40

Solution

In order to use Equations 5.52 and 5.53, we developed Table 5.7 using the values given in Table 5.6.

Table 5.7 **P**rocessed data for use in Equations 5.52 and 5.53

No.	x	y	xy	x^2	y^2
1	10	10	100	100	100
2	15	14	210	225	196
3	18	18	324	324	324
4	20	19	380	400	361
5	23	24	552	529	576
6	25	26	650	625	676
7	27	28	756	729	784
8	30	34	1020	900	1156
9	33	38	1254	1089	1444
10	35	40	1400	1225	1600

$m = 10$ $\sum_{i=1}^{m} x_i = 236$ $\sum_{i=1}^{m} y_i = 251$ $\sum_{i=1}^{m} x_i y i = 6646$ $\sum_{i=1}^{m} x_i^2 = 6146$ $\sum_{i=1}^{m} y_i^2 = 7217$

Substituting the resulting values of Table 5.7 into Equations 5.52 and 5.53, we get

$$\theta = \frac{(10)(6646) - (236)(251)}{5764} = 1.2533$$

and

$$k = \frac{(251)(6146) - (236)(6646)}{5764} = -4.4778$$

Thus, the least-squares line is

$$y = (1.2533)x - 4.4778 \qquad\qquad \textbf{[5.55]}$$

5.9 PROBLEMS

1. Define the terms mean, median, and mode.

2. What are the important axioms of probability?

3. What is the difference between discrete and continuous random variables?

4. The sampled tolerances of a mechanical part, in millimeters (mm), are given in Table 5.8. Calculate the standard deviation and variance of the sample.

Table 5.8   **T**olerances of a mechanical part

Observation No.	Tolerance (mm)
1	0.07
2	0.09
3	0.01
4	0.06
5	0.07
6	0.06
7	0.03
8	0.07
9	0.05
10	0.06
11	0.04
12	0.05

5. Obtain an expression for the Poisson distribution, using the binomial distribution.

6. In an engineering company 10 percent of the parts manufactured are out of specified tolerances. Compute the probability of having 2 rejected parts out of a sample of 30 parts.

7. Prove that the total area under the curve for the exponential probability density function is equal to unity.

8. Prove that the cumulative distribution function associated with the exponential distribution is given by

$$F(t) = 1 - e^{-\lambda t} \qquad\qquad [5.56]$$

where

$F(t)$ is the cumulative distribution function.

λ is the distribution parameter.

t is the time variable.

9. What are the statistical distributions of the two special cases of the Weibull distribution?

10. Table 5.9 presents 25 times-to-failure of an electric motor. Using the Bartlett test, determine if the data are representative of an exponential distribution.

Table 5.9   Times-to-failure (hours)

800	1000	750	350	350
700	250	650	400	400
700	500	800	500	300
400	1000	850	525	600
500	800	550	675	550

11. A company tested 25 identical electric motors, beginning at time $t = 0$ and continuing until the occurrence of the 15th failure, replacing each failed motor with a new one. Compute the exponential minimum mean time to failure of the motors, with a 90 percent confidence level.

12. Table 5.10 presents data for variables P and Q of an electrical design. Fit a least-squares line to the data.

Table 5.10   Data for variables P and Q

P	Q	P	Q
20	20	20	18
35	35	15	16
40	37	24	26
30	31	28	30
25	23	34	40

REFERENCES

1. Walton, J.W. *Engineering Design: From Art to Practice.* New York: West Publishing Co., 1991.
2. Eves, H. *An Introduction to the History of Mathematics.* New York: Holt, Rinehart and Winston, 1976.
3. Spiegel, M.R. *Statistics. Schaum's Outline Series.* New York: McGraw-Hill, 1961.
4. Ertas, A.; and J.C. Jones. *The Engineering Design Process.* New York: Wiley, 1993.
5. Ang, A.H.S.; and W.H. Tang. *Probability Concepts in Engineering Planning and Design.* New York: Wiley, 1975.
6. Dhillon, B.S. *Mechanical Reliability: Theory, Models and Applications.* Washington, D.C.: American Institute of Aeronautics and Astronautics, Inc., 1988.
7. Epstein, B. "Tests for the Validity of the Assumption that the Underlying Distribution of Life Is Exponential." *Technometrics* 2, (1960), pp. 83–101 and pp. 167–183.
8. Lamberson, L.R. "An Evaluation and Comparison of Some Tests for the Validity of the Assumption that the Underlying Distribution of Life Is Exponential." *AIIE Transactions* 12, (1974), pp. 327–335.
9. Dhillon, B.S. *Quality Control, Reliability and Engineering Design.* New York: Marcel Dekker, Inc., 1985.
10. Dixon, W.J.; and F.J. Masset. *Introduction to Statistical Analysis.* New York: McGraw-Hill, 1957.
11. Hoog, R.V.; and E.A. Tanis. *Probability and Statistical Interference.* New York: Macmillan, 1983.
12. Kachigan, S.K. *Statistical Analysis.* New York: Radius Press, 1986.
13. *Reliability Engineering,* ed. Von Alven. Englewood Cliffs, N.J.: Prentice-Hall, 1964.
14. Spiegel, M.R. *Theory and Problems of Probability and Statistics. Schaum's Out-line Series.* New York: McGraw-Hill, 1975.

chapter
6

DESIGN QUALITY ASSURANCE

6.1 INTRODUCTION

Quality is an important consideration in new engineering products. Some knowledge of quality assurance principles is essential for people concerned with engineering design. A quality assurance specialist may even be a member of the design team.

The beginning of the quality control discipline could be attributed to the Industrial Revolution [1], with the introduction of labor specialization. However, it was not until World War I, specifically 1916, that C.N. Frazee of Telephone Laboratories applied statistical concepts to inspection problems. In 1924, Walter A. Shewhart of Western Electric developed quality control charts, which today are regarded as a major breakthrough in quality control. The year 1946 witnessed the formation of the American Society for Quality Control, which has played an instrumental role in the further development of the field. Today, the society has thousands of members spread around the globe, and it publishes and sponsors various journals and conferences. Additional information on quality assurance can be found in journals, textbooks, conference proceedings, and other related documents. This

chapter discusses specific design aspects of quality assurance.

6.2 TERMS AND DEFINITIONS

Several quality control terms and their definitions are as follows [2, 3]:

1. **Quality.** Quality may be described in many different ways: meeting an expectation, conforming to requirements, attribute of a product, and so on.

2. **Quality control.** This is a management activity through which the quality of a manufactured product is controlled to avert the production of damaged or defective products.

3. **Control chart.** This is simply a chart containing control limits. More specifically, it is a graphical approach to determining whether a given process is statistically in control [2].

4. **Control limits.** These are the limits given on a control chart to determine whether a sample statistical measure is within bounds.

5. **Sample.** A sample is a group of items or observations taken from a population. The sample

serves as the basis for information with which to make decisions regarding the population.

6. **Inspection.** This is the process of examining an item with respect to given specifications.

7. **Quality management.** Quality management is the totality of functions concerned with the evaluation and achievement of quality.

8. **Quality measure.** This is a quantitative measure of the characteristics and features of an item or service.

6.3   DESIGN QUALITY CONTROL INFORMATION SOURCES

This section lists selected books and related documents, journals and conference proceedings, and organizations concerned with quality control.

6.3.1   BOOKS AND RELATED DOCUMENTS

1. Feigenbaum, A.V. *Total Quality Control.* New York: McGraw-Hill, 1983.
2. *Quality Control Handbook*, ed. J.M. Juran, F.M. Gryna, and R.S. Bingham. New York: McGraw-Hill, 1979.
3. Burgess, J.A. *Design Assurance for Engineers and Managers*. New York: Marcel Dekker, 1984.
4. Grant, E.L.; and R.S. Leavenworth. *Statistical Quality Control*. New York: McGraw-Hill, 1980.
5. *Quality Assurance Requirements for the Design of Nuclear Power Plants*. ANSI Std. N45.2.11. New York: American Society of Mechanical Engineers, 1974.
6. Enrick, N.L. *Quality Control and Reliability*. New York: Industrial Press, 1977.
7. Dhillon, B.S. *Quality Control, Reliability and Engineering Design*. New York: Marcel Dekker, 1985.
8. *Standard for Software Quality Assurance Plans*. ANSI/IEEE Std. 730. New York: Institute of Electrical and Electronics Engineers, 1981.
9. *Reliability Design Qualification and Acceptance Tests*. MIL-Std.–781C. Washington, D.C.: U.S. Department of Defense, 1980.
10. Burgess, J.A. "Assuring the Quality of Design." *Machine Design*, February 1982, pp. 65–69.
11. Evans, J.R.; and W.M. Lindsay. *The Management and Control of Quality*. New York: West Publishing Company, 1989.
12. Banks, J. *Principles of Quality Control*. New York: Wiley, 1984.

13. DelMar, D.; and G.W. Sheldon. *Introduction to Quality Control.* New York: West Publishing Company, 1988.

6.3.2 JOURNALS AND CONFERENCE PROCEEDINGS

1. *Journal of Quality Progress.*
2. *Quality Progress.*
3. *International Journal of Quality and Reliability Management.*
4. *Quality.*
5. Annual Quality Congress Transactions of the American Society for Quality Control.
6. *Quality Review.*
7. Proceedings of the World Quality Congress.
8. Proceedings of the European Organization for Quality Control.

6.3.3 ORGANIZATIONS

1. American Society for Quality Control, 310 West Wisconsin Avenue, Milwaukee, Wisconsin, USA.
2. British Quality Association, 10 Grosvenor Gardens, London, UK.
3. European Organization for Quality Control, Postfach 2613, CH–300l Berne, Switzerland.
4. International Academy for Quality, Hermann Zeller, Albert-Meyer-Strasse 3, D-8038 Grobenzell, Germany.
5. Association for Quality and Participation, 801-B West 8th St., Cincinnati, Ohio, USA.
6. American Society for Nondestructive Testing, 4153 Arlingate Plaza, Columbus, Ohio, USA.

6.4 QUALITY CONTROL ENGINEERING FUNCTIONS

A typical quality control engineering organization performs many functions. These may be divided into six broad groups [4]:

• New design.
• Incoming material.

- General engineering.
- Product evaluation.
- Process engineering.
- Special activities.

The new design activities are associated with: product and process research, tolerances, specifications, initial samples, trial runs, and quality production methods. The incoming material activities pertain to: supplier capability, inspection methods, quality requirements, and documentation and feedback. The general engineering activities are associated with such items as: quality standards, program audits, training and promotion programs, measurement and analytical facilities, discrepant materials, and methods and procedures. The product evaluation activities are related to such items as: sampling plans, auditing, inspection methods, defect classifications, complaints, and inventory evaluations. The process engineering activities are associated with: process controls, troubleshooting, periodic reviews, and capability studies. The special activities include such tasks as: searching and developing new approaches and techniques, and assisting management on various special study projects requiring quality control related inputs.

Organizing a project or company for quality involves many steps. The basic ones are as follows:

1. Identifying quality-related tasks.
2. Assigning responsibilities for their accomplishment.
3. Breaking the overall work down into elements.
4. Outlining the authority and responsibility for each element.
5. Establishing the interrelationships among the elements.

6.5   CONTRACT QUALITY PROGRAM

Planning is the backbone of any quality program and requires careful consideration. A quality program plan should address [5]: contract reviews, process controls, material procurement, calibration requirements, quality records, manufacturing release inspection, quality skill levels, quality audit schedules, design reviews, tooling design and inspection, product inspection, acceptance testing, nonconforming material control, packaging and shipping instructions, final product inspection, and corrective action.

The quality effort begins at the contract stage of an engineering project and calls for thoroughly reviewing the contract document for quality-related factors [5], such as: inspection requirements, warranty needs, acceptance requirements, applicable documents, packaging and shipping requirements, deliverables, and liaison requirements.

Design reviews are an essential component of engineering product development and are generally conducted on a regular basis. The main objective of design reviews is to assure that the product can be effectively manufactured and maintained during its expected life. A quality representative participating in the design reviews can provide guidance on such requrements [5] as: special inspection tools, levels of test and inspection, calibration accuracy, test and inspection skills, packaging and shipping, special processes, special test equipment, nonstandard parts, age-sensitive material, new part vendors and acceptance, and safety.

6.6   PROCUREMENT QUALITY CONTROL

Occasionally, a manufacturer may have to procure parts and materials from outside sources. According to Hayes and Romig [6], United States manufacturers, on average, spend over 50 percent of their sales revenue on the procurement of parts, materials, etc. from other companies. This obviously calls for an effective procurement quality control program; otherwise, the quality of the resulting final product manufactured could be poor. Several of the reasons [1,7] for having a procurement quality control program are: helping parts/materials/subsystems suppliers understand the stated requirements, rating vendor performance and conformance to specifications, evaluating requirement fulfillments, determining the satisfactory specification of contractual and technical needs in the request for proposals, and providing input to the vendor selection process.

6.6.1   PURCHASED PRODUCTS CONTROL

A quality organization must take various measures to control incoming material effectively. A.V. Feigenbaum [8] has provided the following list of guidelines for this effort.

1. Periodically review the effectiveness of the inspection of incoming items.
2. Audit suppliers regularly.
3. Provide satisfactory storage facilities.
4. Develop effective quality measurement procedures mutually acceptable to both users and suppliers.
5. Maintain the effectiveness of devices used in the procurement quality control program.
6. Provide satisfactory facilities for test and inspection at the receiving end.
7. Use statistical methods to analyze the data concerning incoming materials/ items.
8. Provide appropriate material handling equipment and services.
9. Use acceptance sampling tables.
10. Provide procurement quality information to vendors.
11. Provide appropriate training to procurement inspectors.

12. Develop close relationships with suppliers.
13. Provide a system to dispose of nonconforming materials and components promptly.

6.6.2   INCOMING MATERIAL AND PARTS INSPECTION

The primary objective of the incoming inspection is to determine if the newly procured items effectively meet the specified requirements. Specific reasons for conducting receiving-end inspections are:

1. High probability of having defective units in incoming orders.
2. Possibility of receiving unsafe items.
3. Possibility of lower quality of outgoing products because of defective incoming items.

The incoming inspection may take various forms, some of which are as follows [9]:

1. *All* incoming items are inspected with respect to *all* of the desired requirements.
2. The incoming items are only checked to determine if they are identical to the ones ordered.
3. The incoming items are *sampled* to determine their conformance to required specifications. The decision regarding the acceptance or rejection of the incoming lot is made on the basis of the sampling result.

The receiving-end inspectors perform various tasks, including: conduct the required tests on the received items, make acceptance or rejection decisions per the outlined criteria, place the accepted items at their appropriate locations, and complete the paperwork related to the acceptance or rejection.

6.6.3   PROCUREMENT QUALITY CONTROL FORMULAS

This section presents three formulas for measuring inspector accuracy and waste [3] and vendor quality rating [10]. Since the inspectors may be prone to human error, they could inadvertently accept defective items and reject good ones. To minimize the occurrence of such cases, check inspectors are usually used to review the performance of the regular inspectors. The check inspectors reexamine the output (i.e., the accepted and rejected items) of the regular inspectors, as well as their approach to the inspection process. Formulas I and II provide indexes that are useful for measuring the accuracy and waste, respectively, of the regular inspectors.

The vendor rating is useful for determining the satisfactory quality of the purchased items. Formula III is one of the indexes used in industry to rate vendor quality.

Formula I   This formula is concerned with determining the percentage of defects accurately identified by the regular inspector. The equation is as follows:

$$IA = \frac{(D_{dri} - k)(100)}{D_{mri} + (D_{dri} - k)}$$ [6.1]

where

k is the number of defect-free units rejected by the regular inspector, according to the findings of the check inspector.

D_{dri} is the number of defects uncovered by the regular inspector.

D_{mri} is the number of defects overlooked by the regular inspector, according to the findings of the check inspector.

IA is the percentage of defects accurately identified by the regular inspector.

Formula II   This formula is concerned with determining the percentage of defect-free units rejected by the regular inspector. The equation is as follows:

$$N = \frac{k(100)}{\theta - (D_{mri} + D_{dri} - k)}$$ [6.2]

where

N is the percentage of defect-free units rejected by the regular inspector.

θ is the number of units inspected.

Assume that a regular inspector examined a single lot of units and discovered 80 defects. A check inspector was assigned to reexamine all the units in the lot and found $k = 7$ and $D_{mri} = 12$. Compute the percentage of defects correctly identified by the regular inspector.   **Example 6.1**

Solution

Substituting the given data into Equation 6.1 yields

$$IA = \frac{80 - 7}{12 + 80 - 7} = 85.88\%$$

This means the regular inspector correctly discovered 85.88 percent of the defects.

Formula III   This formula [10] is concerned with the vendor quality rating, VQR, which is expressed as

$$VQR = \frac{AL(100)}{RL}$$ [6.3]

where

   RL is the total number of lots received by the buyer.

   AL is the number of lots acceptable to the buyer.

Example 6.2 | Over a period of one year, an engineering manufacturer received 40 shipments of a certain part from a vendor. Each shipment contained an equal number of parts and four shipments were rejected. Determine the vendor quality rating index.

Solution

The number of lots accepted was $(40 - 4) = 36$. Substituting the given data into Equation 6.3 yields

$$VQR = \frac{36}{40}(100) = 90\%$$

This means that the vendor quality rating index is 90 percent.

6.7 DESIGN QUALITY IMPROVEMENT GUIDELINES

There are many steps that the professionals involved in engineering design may take to improve quality. References 11 and 12 list a number of guidelines for design quality improvement:

1. Simplify the assembly and make it foolproof.
2. Reduce the number of parts as much as possible.
3. Eliminate adjustments.
4. Design for robustness.
5. Eliminate or minimize engineering changes on released products.
6. Design for effective testing.
7. Reduce number of different part numbers.
8. Select components that can withstand process operations.
9. Use repeatable and well-understood processes.
10. Lay out the components for reliable process completion.

Steps such as these will lead to better product reliability, a reduction in the assembly error rate, a lower volume of drawings and instructions to control, a higher part yield, better consistency in part quality, a lower degradation of performance with time, less damage to parts, and better serviceability.

   Today, many engineering systems use computer technology to some degree. For the effective operation of such systems, quality is important for both the computer hardware and the software. The environment in which software is developed possesses particular characteristics that may adversely affect its quality. Some of these characteristics [13] are: poorly defined customer objectives, high programmer turnover, cost and time constraints, hardware complexities, outdated support tools, programmer skill level variation, software-naive customers, and a small project staff.

6.8 T~AGUCHI~ M~ETHODS~

The Taguchi methods are based on Genichi Taguchi's [14–16] new philosophy for solving quality problems. A traditional approach for handling quality problems is based on defining quality as conformance to specifications. As an example, if a given tolerance specification was 0.36 ± 0.03, it would be immaterial whether the actual figure was 0.33, 0.36, or 0.39; the specifications would be equally satisfied. However, Taguchi defines quality as product uniformity around a target value. In our example, that value is 0.36, and the actual achieved values closer to the set target would be better than the values further away from the target. This approach advocates modern operational and engineering approaches to the quality problem.

In Japan, the Taguchi methods are widely used [17]. In North America, Xerox Corporation and Bell Laboratories are reported to have experimented with them in the early 1980s.The ITT Corporation conducted over 2,000 case studies related to these methods and reported over $35 million in savings [14].

To advance his new philosophy for quality problem solving, Taguchi proposed the following three-step approach to product and process design:

1. **System design.** This is the application of fundamental scientific and technological principles to the production of a functional design; its primary objective with respect to production processes is to ascertain those manufacturing processes that can produce the item within given tolerances and limits, at the minimum cost.

2. **Parameter design.** This is the investigation to determine the settings that will lower the variation in product or process performance. Some manufacturing process examples involve variations in input voltage, raw materials, and temperatures [18]. Such variations may lead to nonuniformity in the production processes.

3. **Tolerance design.** This is the process of determining the tolerances that will reduce the total product manufacturing and lifespan cost to its minimum level.

6.9 Q~UALITY~ C~ONTROL~ C~HARTS~

Quality control charts are used to analyze discrete or continuous data collected over a period of time. The chart is a graphical tool for assessing the state of control of a process. Control charts were developed by Walter A. Shewhart, Bell Telephone Laboratories, in 1924, to fulfill such functions as: defining a goal for an operation, helping in the accomplishment of that goal, and examining whether or not the goal had been attained. Before a control chart is used, the following basic questions must be answered [4]:

1. How will the sample be chosen, and at what frequency?
2. What are the characteristics to be examined?
3. What are the required gauges or testing devices?
4. What chart will fulfill the objective?
5. What should the sample size be?

The diagram in Figure 6.1 illustrates the general structure of a typical control chart. The chart is composed of a center line (CL), an upper control limit (UCL) and a lower control limit (LCL). Generally, for most applications, the control charts use control limits of plus or minus three standard deviations from the average of the quality characteristic under consideration. The control limits must be chosen such that, for the process to be under control, almost all of the sample points must fall within the limits. If this is indeed the case, the process is within control and no corrective measures are needed. Whenever a sample point falls outside the upper and lower control limits, it signals that the process may be out of its normal state, and an investigation is warranted.

Control charts are used for various reasons, including [19]:

1. To provide information related to process capability.
2. To stop unnecessary process related adjustments.
3. To improve productivity.
4. To provide diagnostic related information.
5. To prevent nonconformity.

The various types of control charts used in quality control work include: C-chart, p-chart, R-chart, and $\bar{x}$-chart. The type of chart used depends on the area of application.

6.9.1 C-CHART

This is a special kind of *attributes control chart* based on the number of defects per unit. The primary use of a C-chart is to control the number of defects per product or item. The areas in which C-charts would be used include [20]: loose solder connections in a wired product, failures in a radar machine per month, defects on a crossbar frame, and flaws in a square foot of cloth.

Equations for the upper and lower control limits can be developed using Poisson's distribution. The mean value $\bar{c}$ is expressed [1] as

$$\bar{c} = \frac{n}{N} \qquad\qquad \textbf{[6.4]}$$

where

n is the total number of defects.

N is the total number of items/products/equipment.

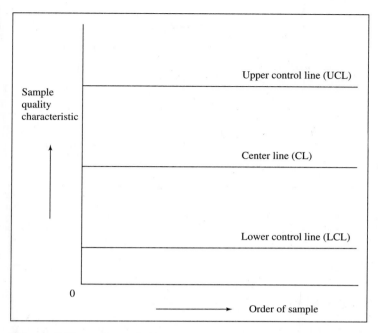

Figure 6.1   A control chart

The standard deviation is

$$\sigma_p = \sqrt{\bar{c}} \qquad\qquad \textbf{[6.5]}$$

These relationships lead to the following equations for the upper and lower control limits, respectively:

$$UCL = \bar{c} + 3\sigma_p \qquad\qquad \textbf{[6.6]}$$

and

$$LCL = \bar{c} - 3\sigma_p \qquad\qquad \textbf{[6.7]}$$

A company manufactures a certain type of computer keyboard and examines its output for defects. The quality control professionals examined 12 keyboards and discovered that keyboards 1, 2, 3, 4, 5, 6, 7, 8, 9, 10, 11, and 12   have   4, 5, 10, 12, 9, 8, 7, 6, 3, 2, 1, and 13 defects, respectively. Develop the C-chart.

Example 6.3

Solution

The total number of keyboard defects is $(4 + 5 + 10 + 12 + 9 + 8 + 7 + 6 + 3 + 2 + 1 + 13) = 80$

Substituting the given data into Equation 6.4, yields

$$\bar{c} = \frac{80}{12} = 6.67 \text{ defects per keyboard}$$

The standard deviation, from Equation 6.5, is

$$\sigma_p = \sqrt{6.67} = 2.58$$

Substituting these results into Equations 6.6 and 6.7, we get

$$UCL = 6.67 + 3(2.58) = 14.41$$

$$LCL = 6.67 - 3(2.58) = -1.08$$

Since the lower control limit, -1.08, is impossible, it is taken as zero. Figure 6.2 shows the C-chart for the given data. Since all of the defects are within the upper and lower control limits, the defect occurrence is normal.

6.9.2 P-Chart

The p-chart is also known as the *percentage chart*. The letter "p" means "proportion," representing the proportion of defective components compared to total components (i.e., percent defective). A p-chart is developed by first collecting 25–30 samples of the attribute measurement under consideration. Examples of such attribute measurements are: the number of defective soldered connections in a piece of wired equipment, the percent defective in the output of a production worker, the number of cracked insulators in a lot, and so on. Note that each sample size should be large enough to have several defective units [12].

Usually, p-charts are used in situations where it is difficult or uneconomical to carry out numerical measurements. The upper and lower control limits are established using the binomial distribution; thus, the mean m and standard deviation σ are as follows

$$m = \frac{n}{Nk} \qquad \text{[6.8]}$$

where

N is the sample size.

k is the number of samples.

n is the total number of defective units in the classification.

and

$$\sigma = \sqrt{\frac{m(1-m)}{N}} \qquad \text{[6.9]}$$

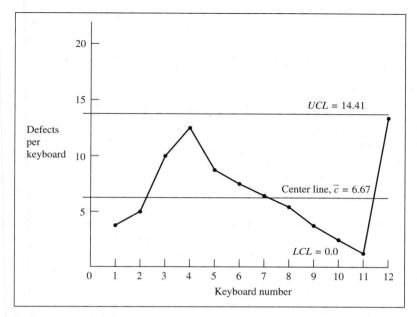

Figure 6.2 The C-chart, for Example 6.3

The upper control limit (UCL) and lower control limit (LCL) are:

$$UCL = m + 3\sigma \qquad \textbf{[6.10]}$$

$$LCL = m - 3\sigma \qquad \textbf{[6.11]}$$

Example 6.4

A production line producing electrical switches was sampled 10 times, with each sample containing 50 switches. After a careful inspection, it was discovered that sample numbers 1, 2, 3, 4, 5, 6, 7, 8, 9, and 10 contained 8, 9, 10, 11, 12, 13, 10, 5, 8, and 7 defectives, respectively. Develop the p-chart.

Solution

The fraction of defectives in the first sample is

$$\frac{8}{50} = 0.16$$

Similarly, the fraction of defectives in samples 2, 3, 4, 5, 6, 7, 8, 9, and 10 are 0.18, 0.2, 0.22, 0.24, 0.26, 0.2, 0.1, 0.16, and 0.14, respectively. Using these data in Equation 6.8 yields

$$m = \frac{93}{(50)(10)} = 0.186$$

Using Equation 6.9 and the available data, we get

$$\sigma = \sqrt{\frac{0.186(1 - 0.186)}{50}} = 0.055$$

Substituting the calculated values into Equations 6.10 and 6.11 leads to

$$UCL = 0.186 + 3(0.055) = 0.351$$

$$LCL = 0.186 - 3(0.055) = 0.021$$

Figure 6.3 shows the p-chart for the given data, clearly indicating that all the sample fractions are within the limits; thus, there is no abnormality in the production line.

6.9.3   R-Chart

The R-chart is the control chart for ranges; it basically shows uniformity or consistency. Narrow and wide R-charts indicate a uniform product and a not-uniform prod-

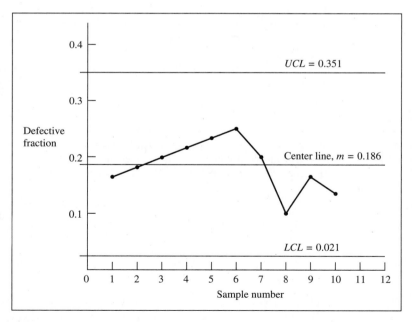

Figure 6.3   P-chart, for Example 6.4

uct, respectively. The R-chart is constructed using the sample ranges as the variability measure. The range is given by

$$SR = SHV - SLV \qquad \textbf{[6.12]}$$

where

SR   is the sample range.

SHV   is the highest observation value in the sample.

SLV   is the lowest observation value in the sample.

According to E.S. Buffa [21], the minimum number of samples for developing an R-chart should be 25, and the number of elements of each sample can be as few as four.

The upper and lower control limits are

$$UCL = D_4(MR) \qquad \textbf{[6.13]}$$

and

$$LCL = D_3(MR) \qquad \textbf{[6.14]}$$

where

MR   is the mean value of the ranges.

D_3 and D_4 are the factors whose values [22] are given in Tables 6.1 and 6.2, respectively.

Table 6.1   Calculated values for the factor D_3

Sample Size	Value for D_3
2	0
3	0
4	0
5	0
6	0
7	0.08
8	0.14
9	0.18
10	0.22
11	0.26
12	0.28
13	0.31
14	0.33
15	0.35

Table 6.2   **C**alculated values
for the factor D_4

Sample Size	Value for D_4
2	3.27
3	2.58
4	2.28
5	2.12
6	2.00
7	1.92
8	1.86
9	1.82
10	1.78
11	1.74
12	1.72
13	1.69
14	1.67
15	1.65

6.9.4 $\bar{X}$-CHART

The $\bar{x}$-chart (pronounced "x-bar chart") is also known as the control chart for averages, where the $\bar{x}$ denotes the average of a sample. This chart belongs to the family of control charts for variables. Examples of such variable measurements are: diameter in inches, resistance in ohms, and length in feet.

The objective of this chart, and the R-chart, is to determine whether the process is statistically in control with respect to variable measurements. As for the R-chart, the minimum number of samples for establishing the $\bar{x}$-chart control limits is 25 [21], and each sample mean value is plotted on the chart. The guidelines for determining the sample size include [23]:

1. Frequently, a sample size of five is taken in the industrial sector, because of simplicity.

2. For a sample size of four or more, the distribution of the sample means approximately follows the normal distribution.

3. An increase in the sample size leads to an increase in the inspection cost per sample.

4. As the sample size increases, the control chart becomes more sensitive to minute variations in the process average value. In fact, the upper and lower control limits become closer to the process average line.

The center line for the chart is given by

$$\bar{\bar{x}} = \frac{\sum_{i=1}^{n} x_i}{n}$$ [6.15]

where

$\bar{\bar{x}}$ is the mean of the sample means, also known as the *grand mean*.

x_i is the mean of sample *i*.

n is the number of samples.

The upper and lower control limits, respectively, are

$$UCL = \bar{\bar{x}} + 3\bar{s}$$ [6.16]

$$LCL = \bar{\bar{x}} - 3\bar{s}$$ [6.17]

where

$\bar{s}$ is the standard deviation of the sample means.

An alternative approach to expressing the upper and lower control limits is as follows:

$$UCL = \bar{\bar{x}} + F(MR)$$ [6.18]

$$LCL = \bar{\bar{x}} - F(MR)$$ [6.19]

The term *F* is defined as

$$F = \frac{3}{d_2\sqrt{N}}$$ [6.20]

where

N is the sample size.

MR is the average of the sample ranges.

d_2 is the factor whose value depends on the sample size. Table 6.3 presents values for d_2.

6.10 TOTAL QUALITY MANAGEMENT (TQM)

The TQM concept is composed of three elements, and the challenge for an organization is to "manage" in such a way that the "total" and the "quality" can and do occur [24]. The objective of TQM is to mobilize all of the possible resources of an organization to satisfy its customers. TQM is a new philosophy developed in the search for the long-term success of an organization.

The beginnings of TQM in the late 1940s may be attributed to such persons as W.E. Deming, J.M. Juran, and A.V. Feigenbaum [25]. After this period, the concept

Table 6.3 Calculated values for the factor d_2

Sample Size	Value for d_2
2	1.13
3	1.69
4	2.06
5	2.33
6	2.53
7	2.07
8	2.85
9	2.97
10	3.08
11	3.17
12	3.26
13	3.34
14	3.41
15	3.47

of TQM was practiced more in Japan than anywhere else. In 1951, the Japanese Union of Scientists and Engineers created a prize named after the quality guru, W.E. Deming, to be awarded to an organization that implemented the most successful quality policies. Subsequently, in 1987, the United States Government introduced a similar award, named the Malcolm Baldrige Award.

The basis for the success of TQM philosophies is the emphasis that all aspects of the company must be effectively managed and all of its employees properly trained, with satisfactory authority to accomplish their jobs effectively. This demands the following four actions from management:

1. **Plan.** This means that management must develop plans for improvement and establish mutually acceptable completion dates with the employees concerned.

2. **Do.** This requires that management implement the above plans.

3. **Check.** This means that management must review the results by specified dates.

4. **Act.** This means that management must take all necessary actions after the review.

6.10.1 TOTAL QUALITY INITIATIVES

Various surveys of a number of companies have revealed many different reasons for embarking on TQM initiatives. The four primary reasons were [24]:

1. **Competition.** Over the years, the competition for goods and services has increased. Many companies have adopted TQM initiatives to increase their chances of keeping their share of the market.

2. **Customer Demand.** The awareness of quality has led many customers to put pressure on companies to provide better quality.

3. **Employees.** The realization of the need to create a work environment that recognizes individual and group contributions was one of the important factors for many companies to implement TQM initiatives.

4. **Internal crisis.** Various internal crises within companies, such as spending a substantial amount of available time to "put out fires," were instrumental in starting TQM initiatives.

6.10.2 BALDRIGE AWARD EVALUATION FACTORS

The Baldrige Award, created by the United States Government to bolster TQM efforts in that country, evaluates candidates on the basis of various core values and concepts. The ten core values and concepts are as follows [26]:

1. **Leadership.** The company leaders' active role in the TQM effort.

2. **Customer-driven quality effort.** The development of approaches and ideas from the point of view of customer satisfaction.

3. **Ongoing improvement.** The continuous effort of the organization with respect to TQM.

4. **Public responsibility.** The attention paid to general business and community issues in the TQM efforts.

5. **Response-time.** The efficiency of responses to customer complaints; the reduction in product production and service introduction cycles; etc.

6. **Partnership development.** The effort spent in promoting internal and external partnerships.

7. **Employee participation.** The degree of participation of company employees in the TQM program.

8. **Design quality.** The effort spent to design quality related measures into the product and services, and the production processes.

9. **Management by fact.** The facts and data used to demonstrate the company's progress toward quality and performance objectives.

10. **Long-range plan.** The long-range plans, strategies, and resource allocations with respect to TQM.

6.10.3 TQM BENEFITS AND PITFALLS

Over the years, companies throughout the world have used the TQM concept and derived various benefits from it. Some of those benefits are as follows [24]:

1. Increase in profit.
2. Reduction in claim efficiency.
3. Improvement in customer satisfaction.
4. Reduction in product failure rate.
5. Increase in sales.
6. Improvement in employee morale.
7. Increase in productivity.
8. Improvement in market share.
9. Reduction in inventory.
10. Reduction in required manpower.

However, the application of the TQM concept is not an automatic success. Many organizations have experienced various difficulties. Some of the more common management-related problems plaguing companies embarking on the TQM concept are as follows [25]:

1. Insufficient allocation of resources for manpower training and development.
2. Failure to delegate decision-making authority to lower levels in the management hierarchy.
3. Management insistence on implementation of the process in a manner the employees find acceptable.
4. Insufficient time devoted to the issues by upper management.

6.11 CASE STUDY: FIRTH OF TAY BRIDGE DISASTER

In December of 1879, the Firth of Tay bridge in Scotland, made up of 200-ft truss spans of wrought iron, collapsed, resulting in 100 deaths [27]. An 80 miles per hour wind blew down thirteen spans, and a train carrying 100 passengers was thrown into the water below, as shown in Figure 6.4. Fowler and Baker, who designed the bridge, conducted many experiments with respect to wind to determine the cause of this failure. Eventually, they correctly concluded that the Tay Bridge disaster was the result of wind pressure.

A subsequent study also revealed that there was no mechanism for testing or approving the materials used in the construction of the bridge. This shortcoming resulted in an extensive use of defective materials. To overcome this shortcoming, regular procedures for material testing and quality control were subsequently developed.

6.12 PROBLEMS

1. Define the following terms:
   a. Quality assurance.

Figure 6.4   The Firth of Tay bridge disaster in 1879.
| The Bettmann Archive

  b. Quality circle.
  c. Control limits.
  d. Control chart.

2. What are the major functions of quality control engineering?

3. Discuss quality related inputs for design reviews.

4. What are the advantages of procurement quality control?

5. Discuss the steps engineering design professionals may take to improve quality.

6. A regular inspector inspected a lot of items and found 70 defects. A check inspector reexamined all the items in the lot (i.e., good plus defective) and discovered that 10 defect-free items were rejected by the regular inspector and 14 defective items were overlooked. Calculate the percentage of defects correctly identified by the regular inspector.

7. Describe the Taguchi approach.

8. Assume that an operator performs a repetitive task. Data on the time required to perform the task are collected randomly nine times over a given period. Each sample includes five observed times, as shown in Table 6.4. Construct the $\bar{x}$-chart and comment on the end result.

Table 6.4   Observed times for samples A to I

| Observation | Sample | | | | | | | | |
No.	A	B	C	D	E	F	G	H	I
1	15	16	21	17	9	18	13	14	18
2	20	25	15	19	15	27	10	13	20
3	18	20	8	20	30	30	22	18	21

Observation	Sample								
No.	A	B	C	D	E	F	G	H	I
4	10	30	31	23	8	14	10	7	10
5	7	11	24	14	14	18	25	17	14

9. Develop the *R*-chart for the data given in Table 6.4 and comment on the end result.

REFERENCES

1. Dhillon, B.S. *Quality Control, Reliability, and Engineering Design*. New York: Marcel Dekker, Inc., 1985.

2. Omdahl, T.P. *Reliability, Availability, and Maintainability (RAM) Dictionary*. Milwaukee, WI: ASQC Quality Press, 1987.

3. *Quality Control Handbook*, ed. J.M. Juran, F.M. Gryna, and R.S. Bingham. New York: McGraw-Hill, 1979.

4. Charbonneau, H.C.; and G.L. Webster. *Industrial Quality Control*. Englewood Cliffs, NJ: Prentice-Hall, 1978.

5. Lesser, D. "Quality Assurance." In *Mechanical Engineers' Handbook*, ed. Myer Kutz. New York: Wiley, 1986, pp. 999–1028.

6. Hayes, G.E.; and H.G. Romig. *Modern Quality Control*. London: Collier-Macmillan Company, 1977.

7. Gage, W.G. "Procurement Quality Planning and Control." *Proceedings of the Annual American Society for Quality Control Conf.*, 1978, pp. 158–161.

8. Feigenbaum, A.V. *Total Quality Control*. New York: McGraw-Hill, 1983.

9. Juran, J.M.; and F.M. Gryna. *Quality Planning and Analysis*. New York: McGraw-Hill, 1980.

10. Lester, R.H.; N.L. Enrick; and H.E. Mottley. *Quality Control for Profit*. New York: Marcel Dekker, Inc., 1985.

11. Daetz, D. "The Effect of Product Design on Product Quality and Product Cost." *Quality Progress*, June 1987, pp. 63–67.

12. Evans, J.R.; and W.M. Lindsay. *The Management and Control of Quality*. New York: West Publishing Co., 1989.

13. Gustafson, G.G.; and R.J. Kerr. "Some Practical Experience with a Software Quality Assurance Program." *Communications of the ACM* 25, (1982) pp. 4–12.

14. Taguchi, G. *Experimental Design* 1. Tokyo: Maruzen Publishing Co., 1976 (in Japanese).

15. Taguchi, G. *Experimental Design* 2. Tokyo: Maruzen Publishing Co., 1977 (in Japanese).

16. Taguchi, G. *Introduction to Quality Engineering*. Tokyo: Asian Productivity Organization, 1986.

17. Ryan, N.E. "Tapping into Taguchi." *Manufacturing Engineering Magazine*, May 1987, pp. 43–46.

18. Taguchi, G.; E.A. El Sayed; and T.C. Hsiang. *Quality Engineering in Production Systems*. New York: McGraw-Hill, 1989.

19. Montgomery, D.C. *Introduction to Statistical Quality Control*. New York: Wiley, 1985.

20. *Statistical Quality Control Handbook*. Indianapolis, IN: AT&T Technologies, 1984.

21. DelMar, D.; and G.W. Sheldon. *Introduction to Quality Control*. New York: West Publishing Co., 1988.

22. Buffa, E.S. *Operations Management: Problems and Models*. New York: Wiley, 1972.

23. Besterfield, D.H. *Quality Control*. Englewood Cliffs, NJ: Prentice-Hall, 1979.

24. Farquhar, C.R.; and C.G. Johnston. *Total Quality Management: A Competitive Imperative*. Report No. 60-90-E,. Ottawa, Ontario, Canada: Conference Board of Canada, 1990.

25. Gevirtz, C.D. *Developing New Products with TQM*. New York: McGraw-Hill, 1994.

26. Hodgetts, R.M. *Blueprints for Continuous Improvement*. New York: American Management Association, 1993.

27. Watson, S.R. "Civil Engineering History gives Valuable Lessons." *Civil Engineering, May 1975.*, pp. 48–51.

chapter

7

DESIGN RELIABILITY

7.1 INTRODUCTION

Reliability engineering helps ensure the success of space missions, maintain the national security, deliver a steady supply of electric power, provide reliable transportation, and so on. Reliability is therefore an important consideration in the design of engineering systems. Reliability may be defined as the probability that an item or piece of equipment will carry out its specified function satisfactorily for the stated period when used under the designed conditions.

The history of the application of probability concepts to electric power generation goes back to the 1930s [1–3]. However, World War II is generally regarded as the beginning of the reliability field, when the Germans introduced the reliability concept to improve their V-l and V-2 rockets [4]. During the period between 1945–1950, the United States Navy, Air Force, and Army conducted various studies on the failure of electronic equipment, equipment repair and maintenance costs, etc. As the result of this effort, in 1950, the Department of Defense formed an ad hoc committee on reliability,

which later became known as the Advisory Group on the Reliability of Electronic Equipment (AGREE). This group published a report in 1957 that led to a specification for the reliability of military electronic equipment. Furthermore, because of the awareness of the reliability problem in the United States, the early 1950s witnessed the appearance of the Institute of Electrical and Electronic Engineers (IEEE) *Transactions on Reliability* and the *Proceedings of the National Symposium on Reliability and Quality Control.* Since the 1950s, thousands of publications have appeared in the field [5,6]. This chapter briefly presents various aspects of reliability engineering.

7.2 SELECTED TERMS, AND THE BATHTUB HAZARD RATE CURVE

Various terms are specific to the field of reliability engineering. Some of the more common terms are as follows.

1. **Hazard rate.** This is the rate of change of the number of failed parts divided by the number of survived parts at time t.
2. **Reliability.** This is the probability that an item will perform its specified function satisfactorily for the stated period when used under the designed conditions.
3. **Maintainability.** This is the probability that a failed item will be repaired to its satisfactory operational state.
4. **Active redundancy.** This term indicates that all redundant items are operating simultaneously.
5. **Failure.** This is the inability of an item to operate within the initially stated guidelines.
6. **Maintenance.** This is all the actions required to keep an item in a defined condition, or restoring it to that condition.
7. **Downtime.** This is the period of time during which an item is not in a condition to perform its stated mission.

The failure behavior of many items may be described by the *bathtub hazard rate curve* shown in Figure 7.1. The curve's resemblance to a bathtub is the reason for its name. The curve can be divided into three parts: decreasing hazard rate region, useful life region, and wear out region.

The decreasing hazard rate region is also known as the burn-in period, infant mortality period, or debugging period. This is the region in which the hazard rate decreases with increasing time. Some of the reasons for the decreasing hazard rate region failures are poor quality control, poor workmanship and substandard parts, poor manufacturing methods, inadequate debugging, and human error.

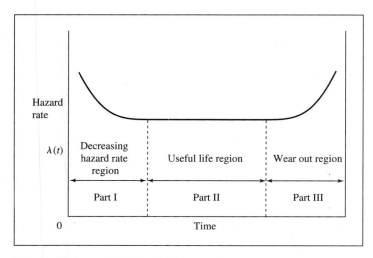

Figure 7.1   Typical bathtub hazard rate curve

In the useful life region, the hazard rate remains constant over time, and the times-to-failure can be represented by an exponential distribution. The failures during this period are assumed to occur randomly and some of the reasons for the failures are low safety factors, abuse, undetectable defects, high unexpected random stress, and natural failures.

The end of the useful life period is the beginning of the wear out region. In this region, the hazard rate increases with increasing time. Some of the causes for wear out region failures are: poor maintenance, incorrect overhaul practices, aging, friction, and corrosion.

7.3 RELIABILITY MEASURES

The common measures used to evaluate the overall reliability of a product are: reliability, mean time to failure (MTTF), and hazard rate. Each of these is discussed in the following sections.

7.3.1 TIME-DEPENDENT RELIABILITY

The *time-dependent reliability* of an item can be obtained using any of the three following Equations 7.1, 7.2, or 7.3 [7]:

$$R(t) = 1 - F(t) = 1 - \int_0^t f(t)dt \qquad \textbf{[7.1]}$$

where

t is time.

$R(t)$ is the time-dependent reliability.

$F(t)$ is the time-dependent probability of failure.

$f(t)$ is the *failure density function* for the item.

or

$$R(t) = \int_t^\infty f(t)dt \qquad \textbf{[7.2]}$$

or

$$R(t) = e^{-\int_0^t \lambda(t)dt} \qquad \textbf{[7.3]}$$

where $\lambda(t)$ is the hazard rate, or the time-dependent failure rate, of the item.

Assume that the times-to-failure of an automobile engine are described by an exponential distribution, as follows:

Example 7.1

$$f(t) = \lambda e^{-\lambda t} \qquad \text{[7.4]}$$

where t is time, and λ is the failure rate of the automobile engine. Obtain an expression for the automobile engine's time-dependent reliability, and the reliability value, for a 50-hour mission, if the engine's failure rate is 0.0005 failures per hour.

Solution

Substituting Equation 7.4 into Equation 7.1 yields

$$R(t) = 1 - \int_0^t \lambda e^{-\lambda t} dt = e^{-\lambda t} \qquad \text{[7.5]}$$

Using the given numerical values in Equation 7.5 leads to

$$R(50) = e^{-(0.0005)(50)} = 0.9753$$

This result means there is 97.53 percent chance that the automobile will not fail during the 50-hour mission.

7.3.2 MEAN TIME TO FAILURE (MTTF)

The *mean time to failure* of an item or system may be obtained using any of the following three relationships:

$$MTTF = \int_0^t t f(t) dt \qquad \text{[7.6]}$$

or

$$MTTF = \int_0^\infty R(t) dt \qquad \text{[7.7]}$$

or

$$MTTF = \lim_{s \to 0} R(s) \qquad \text{[7.8]}$$

where

$\quad s \quad$ is the Laplace transform variable.

$\quad R(s) \quad$ is the Laplace transform of the reliability function.

| Example 7.2 | **C**onsidering Example 7.1, calculate the automobile engine's mean time to failure. |

Solution

Substituting Equation 7.5 into Equation 7.7 yields

$$MTTF = \int_0^\infty e^{-\lambda t} dt = \frac{1}{\lambda} \qquad \textbf{[7.9]}$$

For the given value of λ, from Equation 7.9, the automobile engine's mean time to failure is

$$MTTF = \frac{1}{(0.0005)} = 2,000 \ hours$$

This result means that the engine would fail on an average of every 2000 hours.

7.3.3 HAZARD RATE

The hazard rate of an item or system may be obtained using either of the following two relationships:

$$\lambda(t) = \frac{f(t)}{R(t)} = \frac{f(t)}{1 - F(t)} \qquad \textbf{[7.10]}$$

or

$$\lambda(t) = -\frac{1}{R(t)} \frac{dR(t)}{dt} \qquad \textbf{[7.11]}$$

| Example 7.3 | **O**btain an expression for the hazard rate of the automobile engine discussed in Example 7.1. |

Solution

Substituting Equation 7.5 into Equation 7.11, we get

$$\lambda(t) = -\frac{1}{(e^{-\lambda t})}(-\lambda e^{-\lambda t})$$

$$= \lambda$$

This result means that for exponentially distributed times-to-failure, the engine failure rate is constant.

7.4 BASIC NETWORK RELIABILITY

The components of a system may be interconnected in any one of various forms called *basic networks*. The more common forms are: series, parallel, series–parallel,

parallel–series, k-out-of-m, and standby. This section addresses the reliability evaluation of such networks [4,7].

7.4.1 SERIES NETWORK

A block diagram of this form of network is shown in Figure 7.2. Each block represents a single component of the system. If any one of the components fail, the network fails. In other words, all the components must operate normally for the series network to function successfully.

For independent components, the network reliability, or the probability of success, is

$$P(A_1 A_2 \ldots A_m) = R_{sn} = P(A_1)P(A_2)\ldots P(A_m) \qquad \textbf{[7.12]}$$
$$= R_1 R_2 \ldots R_m$$

where

m is the total number of components in the series system.

A_i is the success event of the component i, for $i = 1, 2, \ldots m$.

$P(A_i) = R_i$ is the probability of success or reliability of component i, for $i = 1, 2, \ldots m$.

$P(A_1 A_2 \ldots A_m)$ is the probability of occurrence of events $A_1, A_2, A_3, \ldots, A_m$.

R_{sn} is the reliability of the series network.

For known component failure rates (i.e., exponentially distributed times-to-failure), the reliability $R_i(t)$ of component i is, using Equations 7.1 and 7.5,

$$R_i(t) = 1 - F_i(t) = 1 - \int_0^t f_i(t)dt \qquad \textbf{[7.13]}$$
$$= 1 - \int_0^t \lambda_i e^{-\lambda_i t} dt$$
$$= e^{-\lambda_i t}$$

where

t is time.

$F_i(t)$ is the failure probability of component i at time t; for $i = 1, 2, \ldots, m$.

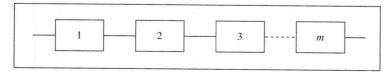

Figure 7.2   A series network

$f_i(t)$ is the failure density function of component i, for $i = 1, 2, \ldots, m$.

λ_i is the constant failure rate of component i, for $i = 1, 2, \ldots, m$.

The time-dependent series network (or system) reliability is obtained by substituting Equation 7.13 into Equation 7.12, as follows:

$$R_{sn}(t) = e^{-\lambda_1 t} e^{-\lambda_2 t} \ldots e^{-\lambda_m t} \qquad \textbf{[7.14]}$$

$$= e^{-\sum_{i=1}^{m} \lambda_i t}$$

To obtain the mean time to failure of the network, we substitute Equation 7.14 into Equation 7.7, which yields

$$MTTF_{sn} = \int_{o}^{\infty} e^{-\sum_{i=1}^{m} \lambda_i t} \, dt \qquad \textbf{[7.15]}$$

$$= \frac{1}{\sum_{i=1}^{m} \lambda_i}$$

where $MTTF_{sn}$ is the mean time to failure of the series network. Substituting Equation 7.14 into Equation 7.11 yields

$$\lambda_{sn}(t) = -\frac{\left(\sum_{i=1}^{m} \lambda_i\right) e^{-\sum_{i=1}^{m} \lambda_i t}}{\left(e^{-\sum_{i=1}^{m} \lambda_i t}\right)} \qquad \textbf{[7.16]}$$

$$= \sum_{i=1}^{m} \lambda_i$$

where $\lambda_{sn}(t)$ is the hazard rate of the series network.

Equation 7.16 is a very important result. It indicates that only in the series case can we sum the failure rates of the system components to obtain the total system failure rate. Usually design specifications call for adding all the failure rates of all parts of a system. This represents the worst-case scenario; that is, if any one of the parts fails, the entire series system fails.

Example 7.4 | **A**ssume that a machine is composed of four independent subsystems acting in series. All the subsystems must function normally for the machine to function successfully. The failure rate of each subsystem is 0.0002 failures per hour. Determine the mean time to failure of the machine.

Solution

Substituting the given data into Equation 7.15, we get

$$\lambda_{sn} = (4\lambda)^{-1}$$
$$= 1250 \; hours$$

This result means that the series system would fail on the average every 1,250 hours.

7.4.2 PARALLEL NETWORK

In a parallel network or system, the active units are connected as shown in Figure 7.3. Each block represents a part or unit of the system. For the success of the system, at least one of the units must function normally. The network is actually a redundant system in that all of the units are active at one time; yet, only one must function normally for system success. This is one of the approaches used to improve system reliability.

The probability of failure F_{sp} of a system with independent units is

$$F_{sp} = P(B_1 B_2 \ldots B_m) = P(B_1)P(B_2) \ldots P(B_m) \qquad \textbf{[7.17]}$$
$$= F_1 F_2 \ldots F_m$$

where

   m is the total number of units or components connected in parallel.

   B_i is the failure event of unit i, for $i = 1, 2, \ldots, m$.

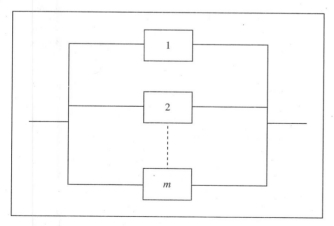

Figure 7.3   A parallel network

$P(B_i) = F_i$ is the probability of failure of unit i, for $i = 1, 2, \ldots, m$.

$P(B_1, B_2, \ldots, B_m)$ is the probability of occurrence of events $B_1, B_2, \ldots, B_m$.

F_{sp} is the failure probability of the parallel network or system.

Subtracting Equation (7.17) from unity, we get the reliability of the parallel network, as follows

$$R_{sp} = 1 - F_1 F_2 \ldots F_m \qquad \textbf{[7.18]}$$
$$= 1 - (1 - R_1)(1 - R2) \ldots (1 - R_m)$$

where R_{sp} is the parallel network reliability.

For a constant failure rate λ_i for each unit, we substitute Equation 7.13 into Equation 7.18 to get

$$R_{sp}(t) = 1 - (1 - e^{-\lambda_1 t})(1 - e^{\lambda_2 t}) \ldots (1 - e^{-\lambda_m t}) \qquad \textbf{[7.19]}$$

where $R_{sp}(t)$ is the parallel network reliability at time t.

For identical units or parts, Equation 7.19 reduces to

$$R_{sp} = 1 - (1 - e^{-\lambda t})^m \qquad \textbf{[7.20]}$$

where λ is the constant failure rate of a unit or part.

Substituting Equation 7.20 into Equation 7.7 yields

$$MTTF_{sp} = \int_o^\infty [1 - (1 - e^{-\lambda t})^m] dt \qquad \textbf{[7.21]}$$
$$= \frac{1}{\lambda} \sum_{i=1}^m \frac{1}{i}$$

where $MTTF_{sp}$ is the mean time to failure of the parallel system.

Example 7.5 | Two independent and identical central processing units in a computer operate simultaneously, and at least one of them must function normally for the successful operation of the computer. Assume that the times-to-failure of the processing units are exponentially distributed, and that the mean value is 1000 hours. Calculate the computer reliability for a 50-hour mission, and its mean time to failure.

Solution

The failure rate λ of a central processing unit is

$$\frac{1}{1000} = 0.001 \text{ failures/hour}$$

Substituting the given data into Equation 7.20 yields

$$R_{sp}(50) = 1 - \{1 - e^{-(0.001)(50)}\}^2$$
$$= 0.9976$$

Similarly, using the known data in Equation 7.21, we get

$$MTTF_{sp} = \frac{1}{0.001}\{1 + 1/2\}$$
$$= 1500 \; hours$$

These results show that the reliability of the computer is 0.9976 and on the average it would fail every 1500 hours.

The reliability analysis of a design under consideration indicates that the reliability of one of the subsystems is 0.75, which is well below the required reliability of 0.95. Determine the number of identical, independent subsystems that must be connected in parallel to achieve the required reliability.

Example 7.6

Solution

For identical and independent units, the parallel network reliability is, from Equation 7.18,

$$R_{sp} = 1 - (1 - R)^m \qquad [7.22]$$

where R is the subsystem reliability. Rearranging Equation 7.22, we get

$$m = \frac{\ln(1 - R_{sp})}{\ln(1 - R)} \qquad [7.23]$$

Substituting the given data into Equation 7.23 yields

$$m = \frac{\ln(1 - 0.95)}{\ln(1 - 0.75)} \simeq 2 \; units$$

This result means that two subsystems, each with a reliability of 0.75, connected in parallel will achieve the required reliability of 0.95.

7.4.3 SERIES–PARALLEL NETWORK

A series–parallel network is the result of combining series and parallel networks such that m parallel subsystems containing k units each are connected in series. When any one of the m subsystems fails, the entire system fails. For independent and identical units or components, the series–parallel network reliability R_{ssp} is given by

$$R_{ssp} = [1 - (1 - R)^k]^m \qquad [7.24]$$

where R is the unit reliability.

For constant failure rates of the units, we substitute Equation 7.5 into Equation 7.24 to obtain an expression for the time-dependent reliability.

$$R_{ssp}(t) = [1 - (1 - e^{-\lambda t})^k]^m \qquad [7.25]$$

Substituting Equation 7.25 into Equation 7.7, we get

$$MTTF_{ssp} = \int_{o}^{\infty} [1 - (1 - e^{-\lambda t})^k]^m \, dt \qquad \textbf{[7.26]}$$

$$= \frac{1}{\lambda} \sum_{i=1}^{m} \left[(-1)^{i+1} \binom{m}{i} \sum_{j=1}^{ik} \frac{1}{j} \right]$$

where $MTTF_{ssp}$ is the mean time to failure of the series–parallel network.

$$\binom{m}{i} \equiv \frac{m!}{i!(m-i)!}$$

Example 7.7 | **A** series–parallel network with independent and identical units is composed of two subsystems in series, and each subsystem contains three parallel units. If the unit failure rate is 0.008 failures per hour, calculate the network reliability for a 150-hour mission.

Solution

Substituting the given data into Equation 7.25 yields

$$R_{ssp}(150) = [1 - (1 - e^{-(0.008)(150)})^3]^2$$

In this case, the series–parallel network reliability is 43.4 percent.

7.4.4 PARALLEL–SERIES NETWORK

A parallel–series network is also the result of combining the series and parallel configurations. However, in this case, m series subsystems containing k units each are connected in parallel. For system success, at least one of the subsystems must function successfully. For independent and identical units, the network reliability is

$$R_{ps} = 1 - (1 - R^k)^m \qquad \textbf{[7.27]}$$

where R_{ps} is the parallel–series network reliability.

For exponentially distributed times-to-failure of the unit, we substitute Equation 7.5 into Equation 7.27 to obtain

$$R_{ps}(t) = 1 - (1 - e^{-k\lambda t})^m \qquad \textbf{[7.28]}$$

Substituting Equation 7.28 into Equation 7.7 then yields

$$MTTF_{ps} = \int_0^\infty [1 - (1 - e^{-k\lambda t})^m]\,dt \qquad \textbf{[7.29]}$$

$$= \frac{1}{k\lambda} \sum_{i=1}^m \frac{1}{i}$$

where $MTTF_{ps}$ is the parallel–series network mean time to failure.

A parallel–series network is composed of two parallel-connected series subsystems, each containing three independent and identical units. The times-to-failure of the units are exponentially distributed, with the mean value at 400 hours. Calculate the network mean time to failure.

Example 7.8

Solution

The failure rate of each unit is

$$\lambda = \frac{1}{400} = 0.0025 \text{ failures/hour}$$

Substituting the given data into Equation 7.29 yields

$$MTTF_{ps} = \frac{1}{3(0.0025)} \sum_{i=1}^2 \frac{1}{i}$$

$$= 200 \; hours$$

Thus, the parallel–series network mean time to failure is 200 hours.

7.4.5 "K-OUT-OF-M" NETWORK

A k-out-of-m network contains a total of m active units connected in parallel, and at least k units must operate normally for the network (or system) to function successfully. The network reduces to a series network for $k = m$ and a parallel network for $k = 1$. For independent and identical units, the network reliability is

$$R_{k/m} = \sum_{j=k}^m \binom{m}{j} R^j [1 - R]^{m-j} \qquad \textbf{[7.30]}$$

$$\binom{m}{j} \equiv \frac{m!}{j!(m-j)!} \qquad \textbf{[7.31]}$$

where $R_{k/m}$ is the k-out-of-m network reliability.

For constant failure rates of the units, we substitute Equation 7.5 into Equation 7.30 to obtain

$$R_{k/m}(t) = \sum_{j=k}^{m} \binom{m}{j} e^{-j\lambda t}[1 - e^{-\lambda t}]^{m-j} \qquad \textbf{[7.32]}$$

Substituting Equation 7.32 into Equation 7.7 leads to

$$MTTF_{k/m} = \int_{0}^{\infty} \left[\sum_{j=k}^{m} \binom{m}{j} e^{-j\lambda t}\{1 - e^{-\lambda t}\}^{m-j} \right] dt \qquad \textbf{[7.33]}$$

$$= \frac{1}{\lambda} \sum_{j=k}^{m} \frac{1}{j}$$

where $MTTF_{k/m}$ is the mean time to failure of the k-out-of-m network.

Example 7.9 | **A**n aircraft has three independent and identical engines, and at least two engines must function normally for the aircraft to fly successfully. Calculate the reliability of the aircraft flying, if the probability of failure of an engine is 5 percent.

Solution

The reliability of an engine is $(1 - 0.05) = 0.95$. Substituting the known values into Equation 7.30, we get

$$R_{2/3} = \sum_{j=2}^{3} \binom{3}{j} R^j (1 - R)^{3-j}$$

$$= \sum_{j=2}^{3} \binom{3}{j} (0.95)^j (0.05)^{3-j}$$

$$= 3(0.95)^2(0.05) + (0.95)^3$$

$$= 0.9928$$

This result shows that, with respect to the engines, the reliability of the aircraft flying is 99.28 percent.

7.4.6   STANDBY NETWORK

A standby network (or system) is an important type of redundancy network in which only one unit is active or operating and m units are on standby (i.e., nonactive). If the operating unit fails, it is immediately replaced by one of the standby's. The system fails when all the standby's and the operating unit fail. If there is one unit operating and one independent and identical unit on standby, and if there is a perfect switch mechanism, the system reliability is [7]

$$R_{sb2} = R(1 - \ln R) \qquad \textbf{[7.34]}$$

where R_{sb} is the two-unit standby system reliability.

For constant failure rates of $(m + 1)$ for the independent and identical units, the reliability of the standby system is

$$R_{sb}(t) = e^{-\lambda t} \sum_{i=0}^{m} (\lambda t)^i / i! \qquad \textbf{[7.35]}$$

The following additional assumptions are associated with Equation 7.35:

1. The standby units are as good as new units.
2. The switching mechanism is perfect.

Substituting Equation 7.35 into Equation 7.7, we get

$$MTTF_{sb} = \int_0^{\infty} \left[e^{-\lambda t} \sum_{i=0}^{m} (\lambda t)i / i! \right] dt \qquad \textbf{[7.36]}$$
$$= \frac{m + 1}{\lambda}$$

where $MTTF_{sb}$ is the mean time to failure of the standby system.

A system has two independent and identical motors, one operating and the other on standby. The motor failure rate is 0.002 failures per hour. As soon as the operating motor fails, the standby is always automatically turned on. If the standby motor remains as good as new in its standby mode, calculate the system reliability for a 100-hour mission. | **Example 7.10**

Solution

Substituting the specified data into Equation 7.35 yields

$$R_{sb}(100) = e^{-(0.002)(100)} \sum_{i=0}^{1} [(0.002)(100)]^i / i!$$
$$= 0.9825$$

This result means that the standby system reliability is 98.25 percent.

7.5 RELIABILITY EVALUATION METHODS

A number of reliability evaluation methods are available to design professionals. Each method has its advantages and disadvantages. The professionals must use their own judgment when choosing one method over another, as a particular method could be more effective than others in a given situation. This section describes four evaluation methods: network reduction, Markov modelling, failure modes and effect analysis (FMEA), and fault trees.

7.5.1   NETWORK REDUCTION METHOD

The network reduction method is a simple, straightforward method of evaluating the reliability of systems composed of series and parallel networks. The approach calls for sequentially reducing the networks to "equivalent" hypothetical units, until the entire network (or system) reduces to a single hypothetical unit. One advantage of this approach is its simplicity; on the other hand, a major disadvantage is its unsuitability for handling components and systems with degraded failure modes. The method is demonstrated by evaluating the reliability of the network shown in Figure 7.4.

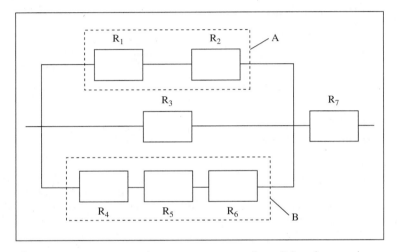

Figure 7.4   A network comprising series and parallel configurations

---

Example 7.11   Evaluate the reliability of the network shown in Figure 7.4, using the network reduction approach. The symbol R_i denotes the reliability of unit i, for $i = 1, 2, 3, 4, 5, 6,$ and 7.

Solution

In this case, the method begins by reducing the subsystems marked A and B to single equivalent hypothetical units. The reliability R_A of subsystem A is, using Equation 7.12,

$$R_A = R_1 R_2 \qquad\qquad \textbf{[7.37]}$$

Similarly, the reliability R_B of subsystem B is

$$R_B = R_4 R_5 R_6 \qquad\qquad \textbf{[7.38]}$$

The network in Figure 7.4 is therefore reduced to the one shown in Figure 7.5. The reliability R_c of subsystem C is, using Equation 7.18,

$$R_c = 1 - (1 - R_A)(1 - R_3)(1 - R_B) \qquad\qquad \textbf{[7.39]}$$

Figure 7.6 shows the reduced network of Figure 7.5. Using Equation 7.12, we obtain the reliability R_n of the network shown in Figure 7.6:

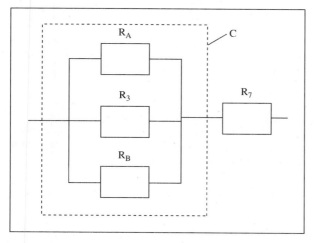

Figure 7.5   The reduced network of Figure 7.4

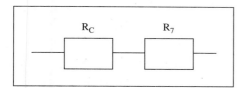

Figure 7.6   The reduced network of Figure 7.5

$$R_n = R_c R_7 \qquad\qquad \textbf{[7.40]}$$

Note that R_n is the reliability of a single equivalent hypothetical unit, or the overall reliability of the network shown in Figure 7.4. This means that the Figure 7.4 network has been reduced to a single equivalent hypothetical unit by using Equations 7.12 and 7.18.

7.5.2 MARKOV MODELLING

Markov modelling is a very powerful approach to evaluating system reliability. The approach is used when the item(s) failure rate is constant, as well as for repairable systems with constant repair rates.

The following assumptions are associated with this technique:

1. All occurrences are independent.
2. The transitional probability of the system going from one state to the next in the finite time interval Δt is given by $(\lambda)(\Delta t)$, where λ is the transition rate (i.e., the constant failure rate or repair rate) from one system state to another.

3. The probability of two or more occurrences in time interval Δt is negligible.

Usually, as the system states increase, the reliability and availability analyses become more difficult. Markov modelling is discussed in detail in References 4 and 7. The method is demonstrated through the following example.

Example 7.12

An air compressor has constant failure and repair rates, and it can only be in either the operating or the failed state. Figure 7.7 shows the state-space diagram of the compressor. Using the Markov approach, develop an expression for the air compressor steady-state availability, and assume that the repaired compressor is as good as new.

Solution

Applying the Markov technique, we obtain the following equations for state 0 and state 1, respectively, of Figure 7.7:

$$P_0(t + \Delta t) = P_0(t)(1 - \lambda_c \Delta t) + P_1(t)\mu_c \Delta t \qquad \textbf{[7.41]}$$

$$P_1(t + \Delta t) = P_1(t)(1 - \mu_c \Delta t) + P_0(t)\lambda_c \Delta t \qquad \textbf{[7.42]}$$

where

t is time.

Δt is the finite time interval.

λ_c is the constant failure rate of the air compressor.

μ_c is the constant repair rate of the air compressor.

$P_0(t + \Delta t)$ is the probability of the air compressor being in operating state 0 at time $(t + \Delta t)$.

$P_1(t + \Delta t)$ is the probability of the air compressor being in failed state 1 at time $(t + \Delta t)$.

$\lambda_c \Delta t$ is the probability of air compressor failure in time interval Δt.

$\mu_c \Delta t$ is the probability of air compressor repair in time interval Δt.

$(1 - \mu_c \Delta t)$ is the probability of no repair in time interval Δt.

$(1 - \lambda_c \Delta t)$ is the probability of no failure in time interval Δt.

$P_i(t)$ is the probability that the air compressor is in state i at time t, for $i = 0, 1$.

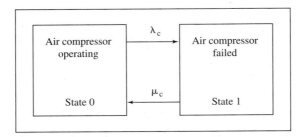

Figure 7.7   Air compressor state-space diagram

In the limiting case, Equations 7.41 and 7.42 become

$$\lim_{\Delta t \to 0} \frac{P_0(t + \Delta t) - P_0(t)}{\Delta t} = \frac{dP_0(t)}{dt} = P_1(t)\mu_c - P_0(t)\lambda_c \qquad \textbf{[7.43]}$$

$$\lim_{\Delta t \to 0} \frac{P_1(t + \Delta t) - P_1(t)}{\Delta t} = \frac{dP_1(t)}{dt} = P_0(t)\lambda_c - P_1(t)\mu_c \qquad \textbf{[7.44]}$$

At time $t = 0$, $P_0(0) = 1$ and $P_1(0) = 0$

To determine the steady-state availability of the air compressor, we set the derivatives of Equations 7.43 and 7.44 equal to zero. Then, utilizing the relationship $P_0 + P_1 = 1$, we obtain

$$P_1\mu_c - P_0\lambda_c = 0 \qquad \textbf{[7.45]}$$

$$P_0\lambda_c - P_1\mu_c = 0 \qquad \textbf{[7.46]}$$

$$P_0 + P_1 = 1 \qquad \textbf{[7.47]}$$

To obtain solutions for P_0 and P_1, discard either Equation 7.45 or Equation 7.46 and solve the remaining two equations. Thus,

$$P_0 = \frac{\mu_c}{\mu_c + \lambda_c} \qquad \textbf{[7.48]}$$

and

$$P_1 = \frac{\lambda_c}{\mu_c + \lambda_c} \qquad \textbf{[7.49]}$$

where

P_0 is the steady-state availability of the air compressor.

P_1 is the steady-state unavailability of the air compressor.

Assume that the constant failure and repair rates of the air compressor are 0.003 failures per hour and 0.004 repairs per hour, respectively. Calculate the compressor's steady-state availability. | **Example 7.13**

Solution

Substituting the given data into Equation 7.48, we get

$$P_0 = \frac{(0.004)}{(0.004) + (0.003)}$$
$$= 0.5714$$

The compressor's steady-state availability is therefore 57.14 percent.

7.5.3 FAILURE MODES AND EFFECTS ANALYSIS (FMEA)

The FMEA method is used to evaluate the reliability of engineering designs. The method was originally proposed in the early 1950s to evaluate the designs of flight control systems [8]. If a critical analysis (CA) is added to the process, the method is known as FMECA. The basic objective of the critical analysis is to rank critical failure-mode effects based on their probability of occurrence. FMEA basically calls for listing all components in a product under design and identifying the possible failure modes of each component, as well as the anticipated failure effects. A comprehensive list of publications on FMEA is given in Reference 9.

Design professionals usually use FMEA starting at the early stage of the design. The basic steps involved are as follows [10]:

1. Define the boundaries and the detailed requirements of the product under design.
2. List all components or parts and the associated subsystems of the product.
3. List all possible failure modes of all components and subsystems.
4. Assign quantitative reliability values to each possible failure mode.
5. Identify each failure-mode effect on all subsystems, the entire product, etc.
6. Enter remarks for each failure mode.
7. Analyze each critical failure mode and initiate corrective measures, as necessary.

The basic characteristics of FMEA include:

1. The process can enhance communication among design professionals.
2. The analysis is a routine upward approach that starts from the component level.
3. Because it evaluates each component's failure effects, the entire design is automatically screened.
4. The analysis points out weak spots in a design and highlights areas for further detailed analysis.

7.5.4 FAULT TREES

A fault tree is a powerful tool for evaluating the reliability of systems during their design phase. The basic difference between FMEA and this method is that FMEA is failure oriented, whereas fault trees are event oriented.

The fault tree method was developed at the Bell Telephone Laboratories in the early 1960s to evaluate the reliability of the Minuteman launch control system. A safety symposium held at the University of Washington, Seattle, in 1965 enhanced the importance of the method. Another symposium held in the middle of the 1970s at the University of California, Berkeley, further increased its importance [11]. The method is described in detail in Reference 12, and a comprehensive bibliography on the topic is given in Reference 6. This section describes the basic aspects of fault trees.

Several symbols are used to construct the fault tree of a complex system [12]. For our purposes, the basic ones are shown in Figure 7.8.

1. **Resultant fault event.** A rectangle denotes a fault event that results from a combination of failure events through the input of a logic gate, such as an AND gate or an OR gate.

2. **Basic fault event.** A circle denotes a basic fault event or the failure of an elementary component. The values of the parameters, such as failure probability, un-

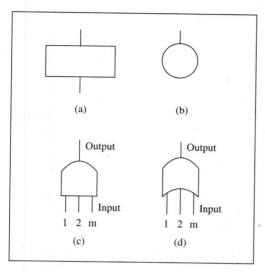

Figure 7.8   Basic fault tree symbols (a) Resultant fault event (b) Basic fault
event (c) AND gate (d) OR gate

availability, failure rate, and repair rate, associated with the basic fault event are obtained from empirical studies or other sources.

3. **AND gate.**   This symbol denotes that an output fault event occurs if *all* of the
input fault events occur.

4. **OR gate.**   This symbol denotes that an output fault event occurs if *one or more*
of the input fault events occur.

The four basic steps involved in developing a fault tree are as follows:

1. Define the top undesired event of the system to be studied.
2. Develop a thorough understanding of the system under consideration.
3. Determine the logical interrelationships of higher-level and lower-level fault
   events.
4. Construct the fault tree using logic symbols.

The development of a fault tree for a simple system is demonstrated through the following example.

Develop a fault tree for a system comprising a windowless room with one switch and three
light bulbs. The switch can only fail to close, and the top undesirable event is the room
without light.   **Example 7.14**

Solution

The fault tree for having a dark room is shown in Figure 7.9. Note that the event "fuseboard
without power" could have been further investigated; however, it was decided to represent

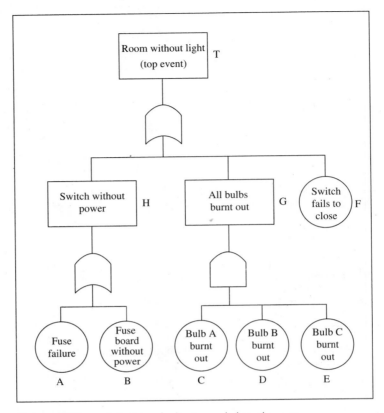

Figure 7.9   A fault tree for having no light in the room

that event with a circle. Each individual capital letter in the figure denotes the corresponding fault event (e.g., A = Fuse failure).

The output fault occurrence probabilities for OR and AND gates, respectively, are

$$F_{OR} = 1 - \prod_{i=1}^{m}(1 - F_i) \qquad \textbf{[7.50]}$$

where

F_{OR} is the probability of occurrence of the OR gate output fault event.

m is the number of independent input fault events.

F_i is the probability of occurrence of input fault event i, for
$i = 1, 2, 3, \ldots m$.

and

$$F_{AND} = \prod_{i=1}^{m} F_i \qquad \textbf{[7.51]}$$

where F_{AND} is the probability of occurrence of the AND gate output fault event.

For very small (i.e., less than 10 percent) occurrence probabilities of input fault events of the OR gate, Equation 7.50 reduces to

$$F_{OR} = \sum_{i=1}^{m} F_i \qquad\qquad \textbf{[7.52]}$$

Example 7.15

Assume that the probabilities of occurrence of basic fault events A ,B, C, D, E, and F shown in Figure 7.9 are 0.1, 0.12, 0.15, 0.15, 0.15, and 0.08, respectively. Calculate the probability of occurrence of the top event (T), that is, the room without light.

Solution

Substituting the given data for events A and B into Equation 7.50, we get

$$F_H = 1 - (1 - 0.1)(1 - 0.12) = 0.208$$

where F_H is the probability of occurrence of event H.

Similarly, using Equation 7.51 and the given data, we obtain

$$F_G = (0.15)(0.15)(0.15)$$
$$= 0.003375$$

where F_G is the probability of occurrence of event G.

Substituting these results and the given data into Equation 7.50 yields

$$F_T = 1 - (1 - 0.208)(1 - 0.003375)(1 - 0.08)$$
$$= 0.2738$$

Thus, the probability of having a room without light is 27.38 percent. The fault tree containing the corresponding event occurrence probabilities is shown in Figure 7.10.

7.6 FAILURE DATA ANALYSIS AND RELIABILITY ALLOCATION

In evaluating the reliability of an engineering design, the two important steps are: performing a failure data analysis, and allocating appropriate reliability values to the system elements. These steps are described in the following sections.

7.6.1 FAILURE DATA ANALYSIS

In design work, it is very important to analyze available failure data carefully. The failure data analysis is essential for establishing the time-to-failure pattern of the given data. This analysis will determine whether the given data follow exponential, Rayleigh, Weibull, or some other distribution pattern. An incorrect assumption regarding the times-to-failure distribution will lead to incorrect reliability conclusions. For example, if it is assumed that the times-to-failure distribution of an item is normal, it means the failure rate is not constant, and it would be totally incorrect to average the failures over a given period of time.

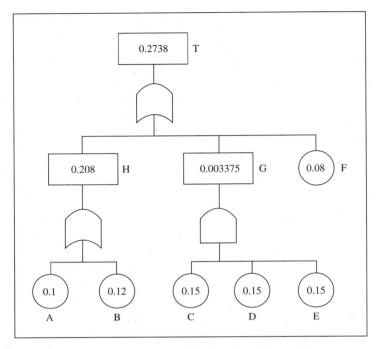

Figure 7.10   A fault tree with event occurrence probabilities

Several techniques are available for performing failure data analyses. One such method is known as *hazard plotting*, which is often preferred by engineers. As its name suggests, the method is very effective in graphically establishing the times-to-failure distribution. This method is described in detail in Reference 13, and some of its advantages are as follows:

1. It fits the times-to-failure data to a straight line.
2. It is equally effective for handling complete and incomplete times-to-failure data.
3. It can be used by people with a limited statistical background.
4. It provides a total, understandable picture of the failure data.

The information that may be obtained from a hazard plot includes:

1. Distribution of the failure data.
2. Values of the representative distribution parameters.
3. The percentage of items failing by a given time period.

The general theory underlying the hazard plot is based on the hazard function

$$\lambda(t) = \frac{f(t)}{1 - F(t)} \qquad [7.53]$$

where

t   is time.

$\lambda(t)$ is the hazard rate function.

$f(t)$ is the times-to-failure density function.

$F(t)$ is the cumulative distribution function.

The cumulative distribution function is defined as

$$F(t) = \int_0^t f(t)dt \qquad [7.54]$$

The cumulative hazard function is defined as

$$\lambda_c(t) = \int_0^t \lambda(t)dt \qquad [7.55]$$

For exponentially distributed failure times, we have

$$f(t) = \lambda e^{-\lambda t} \qquad [7.56]$$

where λ is the distribution parameter. Substituting Equation 7.56 into Equation 7.54 yields

$$F(t) = \int_0^t \lambda e^{-\lambda t} dt \qquad [7.57]$$
$$= 1 - e^{-\lambda t}$$

Using Equations 7.56 and 7.57 in Equation 7.53 leads to

$$\lambda(t) = \frac{\lambda e^{-\lambda t}}{1 - (1 - e^{-\lambda t})} = \lambda \qquad [7.58]$$

Substituting Equation 7.58 into Equation 7.55 results in

$$\lambda_c(t) = \int_0^t \lambda dt = \lambda t \qquad [7.59]$$

Rearranging Equation 7.59, we get

$$t = \frac{1}{\lambda}\lambda_c = \theta\lambda_c \qquad [7.60]$$

where θ is the mean time to failure.

Equation 7.60 is the equation of a straight line passing through the origin. The slope of the line is the estimate for θ. If the plotted data do not yield a straight line, then the failure data do not have an exponential distribution and the sample should be tested for another distribution.

For Weibull distributed times-to-failure, the failure density function is defined by

$$f(t) = \frac{b}{\theta^b} t^{b-1} e^{-(t/\theta)^b} \qquad \theta > 0, b > 0, t \geq 0 \qquad \textbf{[7.61]}$$

where

θ is the scale parameter.

b is the shape parameter.

t is time.

Substituting Equation 7.61 into Equation 7.54 leads to

$$F(t) = \int_0^t \frac{b}{\theta^b} t^{b-1} e^{-(t/\theta)^b} dt \qquad \textbf{[7.62]}$$
$$= 1 - e^{-(t/\theta)^b}$$

Using Equations 7.61 and 7.62 in Equation 7.53 results in

$$\lambda(t) = \frac{b}{\theta^b} t^{b-1} \qquad \textbf{[7.63]}$$

Substituting Equation 7.63 into Equation 7.55 yields

$$\lambda_c(t) = \int_0^t \frac{b}{\theta^b} t^{b-1} dt \qquad \textbf{[7.64]}$$
$$= \left(\frac{t}{\theta}\right)^b$$

Rearranging Equation 7.64, we get

$$t = \theta \lambda_c^{1/b} \qquad \textbf{[7.65]}$$

Taking the natural logarithm of Equation 7.65 leads to

$$\ln t = \ln \theta + \frac{1}{b} \ln \lambda_c \qquad \textbf{[7.66]}$$

The plot of $\ln t$ versus $\ln \lambda_c$ follows a straight line with a slope equal to $1/b$; at $\lambda_c = 1$, the value of θ is equal to time t. In this case, if the plot of the failure data is roughly linear, then the data are representative of the Weibull distribution. The designer can then estimate the parameter values through the given equations. As previously stated, at $b = 1$ and $b = 2$, the Weibull distribution becomes the exponential distribution and the Rayleigh distribution, respectively. Also, if the data plot is not linear, then another distribution must be considered. Equations for other distributions may be developed in similar manner [4].

7.6.2 RELIABILITY ALLOCATION

Once the reliability or related parameter values are defined for a given system being designed, the next logical step is to allocate these overall values down to the component level. The objective is to satisfy the allocated values in the designs of the components or subsystems. In theory, if the allocated reliability values of the components are satisfied, the specified values of the overall system should automatically be achieved. This process is known as *reliability allocation*. Many reliability allocation methods have been developed and are described in Reference 10. This section discusses three of them [12].

Method I The similar familiar systems reliability allocation approach is based on the designer's familiarity with similar systems or subsystems. The method's main weakness is the assumption of adequate reliability and life-cycle cost data for those similar designs.

Method II For the factors of influence method, the allocation is based on four influencing factors: complexity/time, operational environments, failure criticality, and state of the art. Complexity is concerned with the complexity of a component, such as the number of parts; and the time factor deals with the relative operational time of a component during the total functional period of the system. The operational environment factor is concerned with the susceptibility or exposure of components to vibration, high temperature, humidity, etc. The failure criticality factor takes into consideration the critical effect of the failure of a component on the system. The state-of-the art factor is concerned with advancements for the component.

In applying this approach, every item is rated according to its degree of influence by assigning it a number between 1 and 10 (1 means least affected and 10 means most affected). The allocation then uses these assigned numbers to weight all factors.

Method III Combining the previous two approaches results in the combined familiar systems and factor of influence method. This method takes into consideration both the designer's familiarity with similar systems and the influencing factors. This approach can be further strengthened by assigning a certain weight to the in-house failure data, if available, during the allocation process.

7.7 MECHANICAL FAILURE MODES AND SAFETY FACTORS

A failure mode is a physical process or a combination of processes whose effects induce failure. The modes of mechanical failure include: corrosion, fretting, buckling, fatigue, wear, impact, brinelling, creep, stress rupture, ductile rupture, thermal shock, seizure, etc. [7]. Examples of failure modes and associated systems are as follows [14, 15]:

1. **Structural.** Structural failure modes are: fracture, and excessive deflection.

2. **Thermodynamic.** Two associated failure modes are: overheating, and reduction in efficiency.

3. **Fluid.** The associated failure modes are: leakage, and distorted flow.

4. **Kinematic.** Two associated failure modes are: bearing seizure, and reduction in the accuracy of relative movement.

5. **Hydraulic actuation.** The failure modes include: fitting leakage, static seal leak, fluid dirt contamination, and actuator cylinder rupture.

The parameter used to ascertain the safeness of a member or component is called the safety factor. Safety factors based on past experience may provide a satisfactory design; however, a design based solely on such factors could be misleading. There are many ways to define a safety factor [7]. Two of them are described in the following paragraphs.

Definition I   The safety factor S_f is defined as [16]

$$S_f = \frac{Maximum\ safe\ load}{Normal\ service\ load} \qquad \textbf{[7.67]}$$

Generally, this is a good definition, particularly in situations in which the loads are normally distributed.

Definition II   In this case, the safety factor S_f is expressed as [16, 17]

$$S_f = \frac{Mean\ strength}{Mean\ load} \geq 1 \qquad \textbf{[7.68]}$$

This safety factor is generally a good measure in situations in which both the strength and the load are described by a normal distribution. Note that for large variations in strength and/or load, the safety factor becomes meaningless, because the failure rate is positive.

7.8 ENGINEERING DESIGN RELIABILITY GUIDELINES

Over the years, several design reliability guidelines have been developed. These guidelines directly or indirectly contribute to achieving high reliability in the systems being designed. Some of these guidelines are as follows:

1. Test new designs carefully; devise effective test procedures.

2. Simplify a design as much as possible.

3. Use standard parts as much as possible.

4. Use well-tried parts and materials.

5. Allow for human error.

6. Use safety margins of three to six standard deviations (if possible) for critical parameters.

7. Do not use cost-saving approaches at the expense of reliability.

8. Evaluate maintenance with respect to reliability.

9. Use redundancy if required.

10. Evaluate manufacturing with respect to reliability.

11. Consider the effects of transportation, handling, and storage.

12. Carefully analyze all data received from the field.

13. Carefully consider the design of critical parts.

14. Incorporate inspection facilities.

15. Carefully consider available field data when making modifications and improvements to a design.

7.9 CASE STUDY: SPACE SHUTTLE CHALLENGER ACCIDENT

In the early part of 1972, the National Aeronautics and Space Administration (NASA) estimated that to develop and test a space shuttle system would cost $6.2 billion. In August of that year, NASA awarded a contract to Rockwell International Corporation to design and develop the space shuttle orbiter. Martin Marietta, Morton Thiokol Corporation, and Rocketdyne were assigned to produce the external tank, solid rocket boosters, and orbiter main engines, respectively [19].

In April of 1981, the shuttle's ability to go into orbit and return safely was demonstrated by the space shuttle Columbia.

The space shuttle Challenger, Figure 7.11, began its flight on January 28, 1986, at 11:38 AM. (Eastern standard time). The flight ended abruptly 73 seconds later in an explosive burn of the hydrogen and oxygen propellants, destroying the external tank and causing the ultimate structural breakup of the shuttle. As a result of this debacle, President Ronald Reagan established a 13-member commission on the space shuttle Challenger accident. The commission released its report on June 6, 1986.

The report concluded that the Challenger accident was the result of the malfunction of the pressure seal in the right solid rocket motor's aft field joint. The malfunction was caused by a faulty design unacceptably sensitive to many factors: physical dimensions, processing, reusability effects, dynamic loading, and temperature effects.

As corrective measures, the commission proposed nine recommendations concerned with design, shuttle management structure, critical review and hazard analysis, safety organization, communications, landing safety, launch abort and crew escape, flight rate, and maintenance safeguards. Among other things, the design recommendation included replacement of the faulty solid rocket motor joint and seal, either by a new design or by a redesign of the current joint and seal.

Figure 7.11
| UPI/Bettmann

The space shuttle Challenger lifts off from Kennedy Space Center in the early morning of January 28, 1986.

7.10 PROBLEMS

1. Write an essay on the history of reliability engineering.
2. Describe the following terms:
   a. Bathtub hazard rate curve.
   b. Failure.
   c. System availability.
   d. Reliability.
3. What are the reasons for the wear out region of a bathtub hazard rate curve?
4. Prove that the failure probability plus the reliability of a system is equal to unity when the times-to-failure of the system are Weibull distributed.
5. Plot the reliability and failure probability of a product when its times-to-failure are described by an exponential distribution.
6. What are the special-case distributions of the Weibull distribution?

7. An aircraft has four active, independent, and identical engines. At least two of the engines must work successfully for the aircraft to fly. Calculate the reliability of the aircraft with respect to engines, if each engine's probability of failure is 0.1.

8. Assume that in Problem 7, the constant failure rate of an engine is 0.005 failures/hour. Calculate the reliability of the aircraft with respect to engines for a 15-hour flying mission.

9. Obtain a hazard rate expression for an independent-unit series system when failure rate of the units is constant.

10. A computer has two nonidentical and independent central processing units (CPUs) in active redundancy. If the times-to-failure of each CPU are exponentially distributed, obtain an expression for the parallel system mean time to failure.

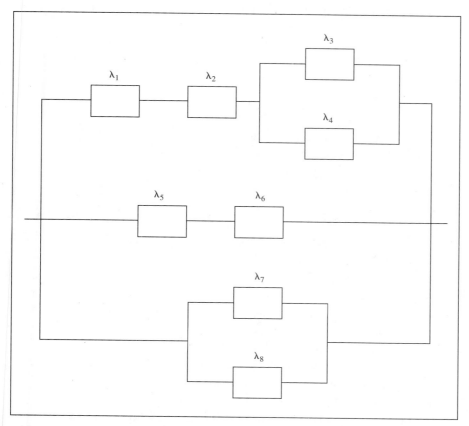

Figure 7.12   A network of eight independent units

11. Obtain an expression for the mean time to failure of the independent-unit network shown in Figure 7.12. Each block in the figure denotes a unit with a constant failure rate of λ_i, for $i = 1, 2, 3, 4, 5, 6, 7, 8$.

12. Discuss the advantages and disadvantages of the fault tree method.

13. Using the Markov method, prove that the unavailability of a system is given by

$$UA(t) = \frac{\lambda}{\lambda + \mu} - \frac{\lambda}{\lambda + \mu} e^{-(\lambda + \mu)t} \qquad \textbf{[7.69]}$$

where

$UA(t)$ is the system unavailability at time t.

λ is the system constant failure rate.

μ is the system constant repair rate.

14. A batch of identical bearings was life tested, and their times-to-failure were 112, 138, 163, 18, 30, 45, 290, 330, 400, 70, 85, 510, 760, 200, and 240 hours. Using the hazard plotting method, determine if the times-to-failure are exponentially distributed.

REFERENCES

1. Smith, S. A. "Service Reliability Measured by Probabilities of Outage." *Electrical World* 103 (1934), pp. 371–374.

2. Layman, W. J. "Fundamental Consideration in Preparing a Master System Plan." *Electrical World* 101 (1933), pp. 778–792.

3. Benner, P. E. "The Use of the Theory of Probability to Determine Spare Capacity." *General Electric Review* 37 (1934), pp. 345–348.

4. Dhillon, B. S. *Reliability Engineering in Systems Design and Operation.* New York: Van Nostrand Reinhold, 1983.

5. Dhillon, B. S. *Reliability Engineering Applications: Bibliography on Important Application Areas.* Gloucester: Beta Publishers, Inc., 1992.

6. Dhillon, B. S. *Reliability and Quality Control: Bibliography on General and Specialized Areas.* Gloucester: Beta Publishers, Inc., 1992.

7. Dhillon, B. S. *Mechanical Reliability: Theory, Models and Applications.* Washington, D.C.: American Institute of Aeronautics and Astronautics, Inc., 1988.

8. Countinho, J. S. "Failure Effect Analysis." *Transactions of the New York Academy of Sciences* 26 (1964), pp. 564–584.

9. Dhillon, B. S. "Failure Modes and Effects Analysis-Bibliography." *Microelectronics and Reliability* 32 (1992), pp. 719–731.

10. Dhillon, B. S. *Systems Reliability, Maintainability and Management.* New York: Petrocelli Books, Inc., 1983.

11. *Reliability and Fault Tree Analysis,* ed. R. E. Barlow, J. B. Fussell, and N. D. Singpurwalla. Philadelphia, PA: Society of Industrial and Applied Mathematics (SIAM), 1975.

12. Dhillon, B. S.; and C. Singh. *Engineering Reliability: New Techniques and Applications.* New York: Wiley, 1981.

13. Dhillon, B. S. *Robot Reliability and Safety.* New York: Springer-Verlag, 1991.

14. McCrory, R. J. "Elements of Realism in Mechanical Reliability." *Proceedings of the American Society of Mechanical Engineers (ASME) Design Engineering Conference–Mechanical Reliability Concepts,* 1965, pp. 1–16.

15. Martin, P. "Reliability in Mechanical Design and Production." *Proceedings of Generic Techniques in Systems Reliabililily Assessment.* Amsterdam: Noordhoff-Leyden Publishers, 1976, pp. 267–271.

16. Phelan, M. R. *Fundamentals of Machine Design.* New York: McGraw-Hill, 1962.

17. Bompass-Smith, J. H. *Mechanical Survival: The use of reliability data.* London: McGraw-Hill, 1973.

18. McCalley, R .B. "Nomogram for Selection of Safety Factors." *Design News,* September 1957, pp. 138–141.

19. *Report of the Presidential Commission on the Space Shuttle Challenger Accident,* vol. 1, June 6, 1986, Washington, D.C.

chapter

8

MAINTAINABILITY DESIGN CONSIDERATIONS

8.1 INTRODUCTION

Today's engineering systems are becoming more and more complex, and greater emphasis is being placed on product maintainability during the product design phase. *Maintainability* is a characteristic that reflects the accuracy, safety, cost effectiveness, ease, and time required to perform any needed maintenance tasks. As an engineering discipline, maintainability may not be as old as reliability, but maintainability concerns have been around for a long time. For example, the Wright brothers' airplane development contract signed with the Army Signal Corps included a clause that the airplane be "simple to operate and maintain" [1]. However, the first United States Department of Defense specification document on maintainability, MIL-M–26512 (USAF), did not appear until 1959. Some of the other related military documents that subsequently appeared were: MIL-STD–778 (definitions of maintainability terms), MIL-STD–470 (maintainability program requirements), MIL-STD–471 (maintainability demonstration), MIL-STD–472 (maintainability prediction), AMCP 706–134 (maintainability guide for design), and AMCP 706–133 (maintainability engineering

theory and practice). This chapter discusses various aspects of maintainability.

8.2 MAINTAINABILITY OBJECTIVES

The reasons for applying maintainability engineering principles [2] include the following:

1. Reducing projected maintenance time and cost through design modifications aimed at ease of maintenance.

2. Using both maintainability and reliability data to determine product availability/unavailability.

3. Determining projected maintenance downtime and comparing this with the allowable downtime, to determine if redundancy is warranted to provide an acceptable level of a required vital function.

4. Determining the amount of labor-hours and related resources needed to carry out the projected maintenance.

Occasionally, the stated (maintainability) requirements are incomplete or unclear. To use this information effectively during the design phase, the maintainability professional must seek answers to such questions as [1]:

1. What is the reason for designing this system?
2. What are the maintenance objectives (e.g., cost, system effectiveness, and maintenance concepts, etc.)?
3. How and where will the product be supported?
4. Are there any policy and environmental considerations? If yes, what are they?
5. What type of manpower is needed to support the product under design?

8.3   MAINTAINABILITY DESIGN CHARACTERISTICS

During the design phase, several maintainability-related primary equipment characteristics [3] should be carefully considered. These characteristics are described in detail in Reference 4–6. Some of them are: interchangeability, modular design, manpower skills, controls, accessibility, standardization, weight, adjustments and calibrations, training requirements, ease of removal, illumination, safety, manuals, labelling and coding, test equipment, test points, lubrication, manpower need, work environment, displays, tools, openings, cases and covers, size and shape, operability, and test hookups. The most frequently mentioned maintainability design factors or characteristics are: controls, displays, test equipment, test points, accessibility, manuals, cases and covers, safety, labelling and coding, handles and handling, connectors, tools, and mounting and fasteners.

An engineering product's maintainability design characteristics incorporate those factors and features that will help decrease equipment unavailability and downtime. The factors that should be carefully considered are: support cost reduction, maintenance ease, preventive and corrective maintenance tasks, and maintenance and support resources minimization. This last factor includes: special maintenance facilities, manpower skill levels, repair parts, and support equipment. The following paragraphs discuss specific features and their effects on the maintainability aspects of engineering design.

8.3.1   CHECKLISTS

Over the years, professionals have reported a number of advantages to using checklists to improve equipment maintainability, particularly during the design phase. Maintainability professionals can develop checklists on those areas deemed essential, and can then use those checklists to [1]:

1. Review a design during various life-cycle phases of a product.
2. Evaluate the influence of certain maintainability design features.
3. Predict equipment maintainability.

An example of a checklist aimed at reducing maintenance downtimes is as follows:

1. Is there a way to detect faults quickly?
2. Is it possible to replace or repair failed items readily?
3. Is there a mechanism to access failed items rapidly?
4. Is there an established repair/discard/replace policy?
5. Are servicing and inspection intervals maximized?
6. Are the provisions in place for periodic checkout under infeasible automatic monitoring?
7. Are there effective provisions in place for inspection?
8. Is there an automatic monitoring mechanism in place for critical performance parameters?
9. Is the mechanism for automatic fault location feasible?
10. Are the indicators or alarms placed in appropriate locations to assist maintenance personnel in promptly locating and handling failed items?
11. Is it possible to localize significant failures to the affected unit, assembly, or equipment?
12. Are there satisfactory test points, displays, controls, adjustments, and checkout procedures to facilitate calibration, checkout, and alignment of the unit after the repair action?
13. Are there sufficient built-in test provisions to verify that corrective maintenance measures have been taken?

8.3.2 EQUIPMENT PACKAGING

A vital maintainability factor is the manner in which equipment is packaged: item layout, part mounting, ease of parts removal, access, etc. Since equipment parts may be located and packaged in various ways, careful attention should be given to such factors as [1].

1. Standardization and modularization needs.
2. Environmental factors: operating stress, vibration, temperature, etc.
3. Manufacturing requirements.
4. Accessibility requirements.
5. Reliability factors.
6. Safety.
7. Qualities peculiar to each part: clearance requirements, servicing needs, weight, size, test equipment access, etc.

Accessibility   Accessibility is the relative comfort with which an item can be reached for such actions as repair, inspection, service, or replacement. Poor accessibility design may lead to such problems as human error, accidental equipment damage, increased repair time, and injuries. The factors that affect accessibility include: distance to be reached, frequency of access usage, type of maintenance task to be performed, specified time requirements for maintenance actions, types of tools required for maintenance actions, work clearances needed to perform required functions satisfactorily, hazards involved with respect to access use, and installation and packaging of items behind the access. In planning for ease of maintenance, an equipment designer should consider such guidelines as those given in Table 8.1.

Modularization   Modularization is the division of a product into separate physical and functional units to assist removal and replacement. The concept of modularization should be designed into a product whenever feasible, particularly where reduction in personnel training and other benefits would result. The concept of modularization cannot be applied to products with equal benefit. Some of the advantages of modular construction are as follows:

1. Usually means an easily maintainable product.
2. Requires relatively low skill levels and fewer tools.
3. Leads to lower maintenance time and cost.
4. Simplifies new product design, thus shortening design times.
5. Permits the use of fully automated methods to produce standard "building blocks."

8.3.3   STANDARDIZATION

Standardization is an important design feature, imposing limitations on the variety of items that must be used to fulfill the product requirements. It should be emphasized

Table 8.1 Maintenance ease guidelines to be considered during the planning phase

No.	Guideline
1.	Locate each equipment item independently with respect to reaching that item.
2.	Use plug-in modules whenever feasible and economical.
3.	Locate each equipment item in such a way that access to it is not blocked by structural items.
4.	Only one access should need to be used to remove any line replaceable unit.
5.	Do not locate items beneath floor boards, structural members, hoses, pipes, and the operator's seat.
6.	Locate items in such a way that a reasonable amount of room is provided for the use of test probes and other tools.
7.	Aim for item removal through the front of the equipment, rather than through its side or back.

that the concept of standardization should be applied to all phases of design, as well as to parts already in the supply system. Standardization has many goals:

1. Reduce the use of different types of parts.
2. Use the maximum number of common parts in different products.
3. Maximize the use of interchangeable parts.
4. Maximize the application of standard off-the-shelf parts.

The advantages of standardization include: reduced product life-cycle acquisition and support cost, and increased product reliability and maintainability.

8.3.4 INTERCHANGEABILITY

Closely related to standardization, interchangeability means that a given part can be replaced by any like part, and the new part must physically fit where the old part did and must be able to carry out the specified functions of that part effectively. The factors to consider in determining interchangeability requirements are: field conditions, and cost effectiveness of manufacture and inspection. Maximum interchangeability can only be achieved if the design professionals ensure the following [1]:

1. In situations where physical interchangeability is a design characteristic, functional interchangeability also exists.
2. Physical similarities include shape, size, mounting, and other physical characteristics.
3. Physical interchangeability does not exist in places where functional interchangeability is not expected.
4. Adequate information is given in the task instructions and in other relevant areas to allow a decision to be made as to whether or not similar items are interchangeable.
5. There is total interchangeability for all items that are expected to be identical, are identified as being interchangeable, have the identical number assigned by the manufacturer, etc.

8.3.5 HUMAN FACTORS

Maintainability depends on both the operator and the maintainer; hence, human factors are very important. Since the environment in which equipment is to be supported may vary significantly from one application to another, maintainability designers must take environment into account. Some of these environments are as follows:

1. **Physical.** This type of environment includes such factors as: noise, radiation, toxic fumes, temperature, vibration, dust, wind, rainfall, and pressure.

2. **Operational.** This type of environment includes such factors as: ventilation, illumination, maintenance workspace arrangement, work duration, and acoustics.

3. **Human.** This type of environment includes such factors as: psychological, physical, physiological, and human limitations.

One of the most important environmental factors affecting maintainability is illumination. Maintenance requires an adequate level of illumination; otherwise, accuracy or effectiveness could be sacrificed. A variety of factors should be carefully considered when designing a lighting system: lighting uniformity, appropriate brightness, light source or work surface glare, illuminants and surfaces quality and color, and brightness contrast between task and background [1].

According to one aircraft maintenance study conducted over a period of 15 months, human error led to 475 accidents and incidents in flight and ground operations. In addition, the study reported that 95 aircraft were damaged or destroyed, with a loss of 14 lives, and most of the accidents occurred shortly after periodic inspections [1]. The basic causes of these human failures were: poor inspection, poor basic training (in relevant maintenance policies, procedures, and practices), poor maintenance training with respect to equipment involved, and inadequate supervision. These indicate that, directly or indirectly, maintainability designers must make improvements towards the following primary goals:

1. Designing equipment such that the incorrect performance of a task becomes difficult or impossible.

2. Minimizing the number and frequency of support tasks required.

3. Designing equipment such that the required tasks performed by a person with specified skills are simplified.

8.3.6 SAFETY

Safety is an important maintainability design factor because maintenance personnel may be exposed to hazards and accident situations. The hazards could be due to inadequate attention being given to safety during the equipment design phase. Human safety guidelines include the following:

1. Install fail-safe devices, as appropriate.

2. Fit all access openings with appropriate fillets and rounded edges.

3. Install items requiring maintenance such that danger in accessing them is minimized.

4. Carefully study potential sources of injury by electric shock.

5. Provide adequate fire fighting equipment in the area where the equipment is installed.

6. Provide appropriate emergency doors and other emergency exits for maximum accessibility.

7. For toxic materials handling, provide adequate eye baths, showers, and other appropriate first aid equipment.

8. Provide adequate guides, tracks, and stops to facilitate equipment handling.

9. Ensure that the weightlifting or holding capacity of lifts, hoists, stands, etc., is clearly marked.

Study after study has indicated that many accidents happen because of human error. Equipment designers and users must take whatever actions are necessary to minimize the possibility of human error. Guidelines proposed for this very purpose include:

1. Design mating parts such that they can only be put together in the correct form.

2. Inspect all completed work carefully.

3. Provide adequate training for all concerned persons.

4. Develop effective support procedures to reduce the occurrence of human error to a minimum.

5. Develop mechanisms to ensure that the established support procedures are being properly followed.

6. Educate each involved individual regarding the possible consequences of performing their assigned tasks incorrectly.

7. Ensure that the proper tools are provided to and used effectively by maintenance personnel.

8. Adjust safety equipment as appropriate.

9. Make everyone safety conscious.

8.4   GENERAL MAINTENANCE DESIGN GUIDELINES

Maintenance oriented design guidelines should be tailored to the specific equipment being developed. However, there are many general maintenance areas that should be addressed by every design [8]: accessibility, environment, test points, lubrication, tools and test equipment, covers and panels, cables, connectors, maintenance minimization, safety, etc. Each of these areas is briefly described in the following paragraphs.

1. **Accessibility.** This includes such factors as: adequate illumination and workspace, adequate sized openings, fewer and easy-to-operate fasteners, and appropriate access to high maintenance frequency areas.

2. **Environment.** This includes protection of equipment from such factors as: hot and cold temperatures, electrical surges, static, high and low humidity, and pressure.

3. **Test points.** This includes such factors as: adequate illumination, clear marking, functional grouping, physical damage protection, and common test equipment accessibility.

4. **Lubrication.** This includes such factors as: freedom from disassembly (breakage), inclusion of mechanism to detect possible future damage, and sealing of bearings and motors.

5. **Tools and test equipment.** This includes such factors as: standardization, metric (if applicable) compatibility, and minimum number required.

6. **Covers and panels.** This includes such factors as: easy opening and replacing, intrusion protection, and practical finishes and color.

7. **Maintenance minimization.** This includes such factors as: fault tolerant design, high reliability, and lifetime components usage.

8. **Cables.** This includes such factors as: individual wire identification, appropriate clamping, and removeable units fabrication.

9. **Connectors.** This includes such factors as: misconnection elemination, efficient disconnect action, and keyed alignment.

8.5   MAINTAINABILITY MEASURES

In maintainability analyses, various types of measures are used, including mean time to repair, mean corrective maintenance time, mean time to detect, and fault detection probability. These measures are described in the following sections.

8.5.1   MEAN TIME TO REPAIR (MTTR)

The MTTR is probably the most commonly used maintainability measure. It measures the elapsed time needed to carry out a maintenance task. It is also used to determine equipment availability and downtime. The MTTR is expressed as [9]

$$MTTR = \frac{\sum_{i=1}^{m} \lambda_i RT_i}{\sum_{i=1}^{m} \lambda_i} \qquad \textbf{[8.1]}$$

where

RT_i is the repair time of item i.

m is the total number of items.

λ_i is the constant failure rate of item i.

Note that the MTTR is the average of the number of times to repair weighted by the occurrence probability.

8.5.2   MEDIAN CORRECTIVE MAINTENANCE TIME (MCMT)

The MCMT is useful for determining the time within which 50 percent of all corrective maintenance actions are to be accomplished. Its estimation depends on equipment characteristics determining the most appropriate probability distribution representing the times to repair. Thus, for the log-normal distribution, the median corrective maintenance time (MCMT) is given by [10]

$$MCMT = antilog \left[\frac{\sum_{i=1}^{m} \log RT_i}{m} \right] \qquad \textbf{[8.2]}$$

where

$\log RT_i$ is the logarithm of RT_i.

m is the total number of items.

8.5.3   MEDIAN PREVENTIVE MAINTENANCE TIME (MPMT)

The MPMT is used to determine the time within which 50 percent of all preventive maintenance tasks are to be performed. For the log-normal distributed preventive maintenance times, the median preventive maintenance time (MPMT) is [10]

$$MPMT = antilog \left[\frac{\sum_{i=1}^{m} \log PT_i}{m} \right] \qquad \textbf{[8.3]}$$

where

m is the total number of items.

$\log PT_i$ is the logarithm of the preventive maintenance time of item i.

8.5.4   MEAN TIME TO DETECT (MTTD)

When an equipment component is dormant and cannot be continuously monitored, periodic tests can be performed to determine if the component is capable of proper operation. Failure may occur during dormancy, and it may only be detectable through such testing. The time between the failure and the test is called the time to detect. The average of these times is the mean time to detect (MTTD) and is expressed as [9]

$$MTTD = \frac{(INT)}{[1 - e^{\{-\lambda(INT)\}}]} - \frac{1}{\lambda} \qquad \textbf{[8.4]}$$

where

λ is the constant failure rate, expressed in failures per hour.

INT is the test interval, expressed in hours.

8.5.5 FAULT DETECTION PROBABILITY

Equipment availability depends upon fault detection accuracy. The probability of correctly detecting faults is given by [9]

$$P_{cf} = \left[\frac{\sum\limits_{i=1}^{m} \sigma_{di}\lambda_{di}}{\sum\limits_{i=1}^{M} \lambda_i} \right] \qquad \textbf{[8.5]}$$

where

P_{cf} is the probability of correctly detecting faults.

λ_i is the constant failure rate of item i.

M is the number of items.

m is the number of items having some degree of fault detectability.

λ_{di} is the constant failure rate of the detectable portion of ith item's constant failure rate.

σ_{di} is that portion of the constant failure rate which is detectable for item i.

8.5.6 PROBABILITY OF REPAIR WITHIN SPECIFIED DOWNTIME

It is not unusual for the design specifications to require repairing a failed component within a given time interval. The probability of performing such a maintenance action may be calculated using the following relationship [9,11] :

$$P(T) = \int_0^T f(t)dt \qquad \textbf{[8.6]}$$

where

$P(T)$ is the probability of performing a maintenance task within a given repair time interval $[0,T]$.

T is the allowable equipment downtime.

$f(t)$ is the equipment repair time probability density function.

t is the repair time variable.

Exponential Distribution For exponentially distributed equipment repair times, we have

$$f(t) = \frac{1}{MTTR} e^{-(\frac{1}{MTTR})t} \qquad \textbf{[8.7]}$$

where

$MTTR$ is the mean time to repair, or the expected downtime, of the equipment.

t is the repair time variable.

Substituting Equation 8.7 into Equation 8.6 yields

$$PT = 1 - e^{\left(-\frac{1}{MTTR}\right)T} \qquad \textbf{[8.8]}$$

The failures, NF, that cannot be repaired within the specified time interval are given by

$$NF = \lambda T_m e^{-\left(\frac{1}{MTTR}\right)T} \qquad \textbf{[8.9]}$$

where

λ is the item constant failure rate, expressed in failures per hour.

λT_m is the mean number of failures in a mission time T_m.

Similarly, the failures, NF_r, that can be repaired within the specified time interval are given by

$$NF_r = \lambda T_m (PT) \qquad \textbf{[8.10]}$$
$$= \lambda T_m \left(1 - e^{-\left(\frac{1}{MTTR}\right)T}\right)$$

Log-Normal Distribution This distribution widely used in maintainability assumes that equipment repair times are log-normally distributed. The repair time probability density function is [1]

$$f(t) = \frac{1}{(t-k)\sigma\sqrt{2\pi}} e^{\left[-\frac{1}{2}\left\{\frac{\ln(t-k)-\mu}{\sigma}\right\}^2\right]} \qquad \textbf{[8.11]}$$

where

k is the constant representing the shortest time in which a repair task can be carried out.

t is the equipment repair time.

μ is the mean of the natural logarithms of the equipment repair times.

σ is one standard deviation of the natural logarithms of the equipment repair times around μ.

Substituting Equation 8.11 into Equation 8.6 yields

$$P(T) = \int_0^T \frac{1}{(t-k)\sigma\sqrt{2\pi}} e^{\left[-\frac{1}{2}\left\{\frac{\ln(t-k)-\mu}{\sigma}\right\}^2\right]} dt \qquad \textbf{[8.12]}$$

Gamma Distribution   The gamma distribution is one of the most flexible distributions for representing various types of repair related data. It is also mathematically tractable. The repair time probability density function is defined as [1]

$$f(t) = \frac{\mu^m}{\Gamma(m)} t^{m-1} e^{-\mu t} \qquad \text{[8.13]}$$

where

μ and m are positive constants (μ represents the scale parameter, and m is the shape parameter).

t is the equipment repair time.

$\Gamma(m)$ is the gamma function as defined by Equation 8.14.

$$\Gamma(m) = \int_0^\infty t^{m-1} e^{-t} dt \qquad \text{[8.14]}$$

For $m = 1$, we get $\Gamma(1) = 1$, and the gamma distribution becomes the exponential distribution, with μ denoting the item's constant repair rate.

Substituting Equation 8.13 into Equation 8.6 yields

$$P(T) = \frac{\mu^m}{\Gamma(m)} \int_0^T t^{m-1} e^{-\mu t} dt \qquad \text{[8.15]}$$

A ssume that the repair times of an aircraft engine are exponentially distributed, with a mean time-to-repair of 5 hours. Calculate the probability of accomplishing a repair within the allowable downtime of 10 hours.

Example 8.1

Solution

Substituting the given data into Equation 8.8 yields

$$P(T) = 1 - e^{-\left(\frac{1}{5}\right)(10)}$$
$$= 0.8646$$

This result means there is an 86.46 percent chance that the engine will be repaired within the allowable downtime.

8.6   CASE STUDY: TETON DAM FAILURE

The failure of Teton Dam (Figure 8.1) in 1976 led to the rural section of eastern Idaho being declared a major disaster area, because almost 300,000 acre–ft of water was dumped into the flatlands downstream of the dam [12]. This disaster resulted in 11 deaths, 2,000 injuries, and the destruction of 7,000 homes and businesses. A total

Fig. 8.1   Aerial view of Teton Dam.

| UPI/Bettmann

estimate for property damage was $1 billion. A 10-member independent panel was subsequently established to investigate the cause of the dam failure. In its final report, the panel concluded that the basic cause was deficiencies in the dam design. The panel specifically pointed out that piping eroded the base or foundation of the embankment's impermeable core material in the keyway. Water burst through the downstream face of the dam because of poor drainage.

The findings of the Interior Department's Internal Review Group (IRG) supported the panel's conclusion, saying that the cause for the dam's collapse was inadequate embankment design and unsatisfactory inspection by the Bureau of Reclamation.

8.7 PROBLEMS

1. List at least five objectives of maintainability.
2. Define "maintainability design characteristics" and list at least 10 of them.
3. Describe the following terms with respect to equipment maintainability:

    a. Accessibility.
    b. Modularization.
    c. Human factors.
    d. Interchangeability.

    e. Safety.

4. Discuss the following design related maintenance factors:

    a. Accessibility.

    b. Tools and test equipment.

    c. Lubrication.

    d. Test points.

    e. Covers and panels.

5. What is the most important use of the median corrective maintenance time?

6. Develop an expression for calculating the probability of repair within the allowable downtime when the equipment repair times are Weibull distributed.

7. An air compressor's times-to-repair are exponentially distributed, with a mean value of 4 hours. Determine the probability of accomplishing a repair within the allowable downtime of 8 hours.

REFERENCES

1. *Maintainability Engineering, Theory and Practice.* AMCP-706-133. document no. ADA-026-006, Springfield, VA: National Technical Information Service (NTIS), 1976.

2. Grant-Ireson, W.; and C.F. Coombs. *Handbook of Reliability Engineering and Management.* New York: McGraw-Hill, 1988.

3. Kline, M.B. *Maintainability Considerations in System Design.* Ph.D. Dissertation, Dept. of Engineering, University of California at Los Angeles (UCLA), Los Angeles, 1966.

4. *Maintainability Guide for Design.* AMCP 706-134. Washington, D.C.: U. S. Department of Defense, 1975.

5. Wohl, J.G. "Why Design for Maintainability?" *IRE Transactions on Human Factors in Electronics* HFE2 (1961), pp. 87–92.

6. Altman, J.W. *Guide to Design of Mechanical Equipment for Maintainability.* report no. ASD-TR-61-381. Dayton, OH:Aeronautical Systems Division, U.S. Air Force, 1966.

7. Rigby, L.V. *Guide to Integrated System Design for Maintainability.* report no. ASD-TR-61-424. Dayton, OH:Aeronautical Systems Division, U.S. Air Force, 1961.

8. Patton, J.D. *Maintainability and Maintenance Management.* Research Triangle Park, NC: Instrument Society of America, 1980.

9. *Handbook of Reliability Engineering and Management,* ed. W. Grant-Ireson, and C.F. Coombs. New York: McGraw-Hill, 1988.

10. Blanchard, B.S. *Logistics Engineering and Management.* Englewood Cliffs, NJ: Prentice-Hall, 1981.

11. Dhillon, B.S. *Reliability Engineering in Systems Design and Operation.* New York: Van Nostrand Reinhold, 1983.

12. Ross, S.S. *Construction Disasters: Design Failures, Causes, and Prevention.* New York: McGraw-Hill, 1984, pp. 148–163.

chapter

9

SAFETY DESIGN CONSIDERATIONS

9.1 INTRODUCTION

One important consideration in engineering design is that the resulting product must be safe for humans. In recent times, the liability decisions by the courts have further increased the importance of safety. A survey of court actions in 1973 [1] highlights several aspects of product liability actions:

1. The basis for 42 percent of the cases reviewed was strict liability.

2. The breach of warranty accounted for 40 percent of the cases.

3. Negligence was the reason for 18 percent of the cases.

Furthermore, according to the National Safety Council [2], over 105,000 Americans were accidentally killed; and over 10 million suffered a disabling injury in 1980 which translated into $83.2 billion. This figure represents over 8 percent of the gross national product of the United States, clearly indicating that safety is a very important factor that should be carefully considered during product design.

The concern for safety is not new; it has been there for thousands of years. For example, Hippocrates identified lead poisoning in the fourth century BC and Pliny the Elder (AD 23–79) pointed to the dust from mercury ore grinding, as well as the fumes from lead, and recommended that workers wear protective masks made from bladder. Over 1400 years later, George Bauer (1492–1555) wrote a 12-volume series on mining and metallurgy and associated hazards. In modern times, one of the earliest safety devices developed was the miner's safety lamp. This was the creation of Humphrey Davy, at the request from the Society for the Prevention of Accidents in Coal Mines [3].

In 1931, H.W. Heinrich published a book on industrial safety [4], and in 1962, the United States Department of Defense developed a military document entitled *System Safety Engineering for the Development of United States Air Force Ballistic Missiles*. Today, many books, journals, conference proceedings, and other documents on safety are available and many organizations are totally devoted to the subject. References 5 and 6 provide information on these areas and Reference 7 lists selected references on safety.

9.2   SYSTEM SAFETY FUNCTIONS

There are many system safety functions including [7]:

1. Developing accident prevention requirements for the basic design.
2. Participating in the design reviews.
3. Developing accident investigation plans.
4. Maintaining accident/safety related data.
5. Performing hazard analyses during the product design cycle.
6. Developing a product/system safety management plan.
7. Participating in accident investigations.
8. Determining emergency procedures.
9. Interacting with safety concerned bodies.
10. Providing training and education on safety.
11. Recommending the needs for safety-related studies.

The National Safety Council [8] has developed many safety-related guidelines for product and process designers. Some of these guidelines are given in Table 9.1.

Table 9.1　Safety related guidelines for designers

No.	Guideline
1.	Eliminate hazards by modifying design, maintenance procedures, or materials used.
2.	Control hazards at their source through guarding, enclosing, or capturing.
3.	Provide necessary warnings/instructions in documentation and display them in effective places.
4.	Educate people to be aware of hazards and to follow appropriate steps to avoid them.
5.	Expect certain abuse and misuse, and take necessary measures to reduce their consequences.
6.	Provide suitable protective equipment to people, and develop steps for ensuring that it is effectively used.

9.3   PRODUCT LIFE-CYCLE SAFETY TASKS

The life cycle of a piece of equipment may be divided into four phases: concept, design and development, manufacture, and operation. Certain safety related tasks are specifically pertinent to each phase. The safety oriented tasks associated with the concept phase include [7, 9]:

1. Develop the equipment/product safety plan.
2. Provide safety-related input to meetings concerned with concept formulation.

3. Develop equipment/product safety information and documentation file.
4. Prepare for preliminary hazard analysis.

Safety effectiveness is determined during the design and development phase. The safety specialist ensures the minimization or elimination of all hazards. The overall safety effort is directed toward such areas as: safety plan implementation, safety requirements incorporation into subcontractors' specifications, regular participation in design meetings, safety related decision documentation, and updating of preliminary hazard analysis.

According to W. Hammer [10], manufacturing defects are second only to design related defects as accident causes. Examples of manufacturing defects are given in Table 9.2. These resulted in recall actions initiated by the United States Consumer Product Safety Commission, as similar defects caused accidents in the past.

Table 9.2 Examples of manufacturing defects

No.	Product	Defect that May Cause Hazard
1.	Gas furnace	Furnace heat exchanger cracks may cause leakage of carbon monoxide into the heated air stream.
2.	Snowmobile	Defective welding may lead to breakage of steering handle.
3.	Aluminium electric fry pan	Heating element crack may cause hazard.
4.	Wringer-washer	Misapplied wire connectors on motor cord may cause electric shock.
5.	Television sets (color)	Line cord inspection indicates manufacturing related damage that may lead to shock hazard.

Safety steps during the manufacturing phase include [7]:

1. Review and evaluate processes and test procedures with respect to safety.
2. Perform (or review, if already accomplished in earlier phases) an operational system safety analysis.
3. Verify tests and inspections to be performed by quality control personnel.
4. Develop failure feedback system to review failures from the point of view of safety.
5. Develop mechanisms to ensure that the safety related actions are performed according to specifications.

In the operation phase, a result of the safety effort in earlier phases will become apparent. In fact, if the safety related measures of preceding phases were effective, this phase should be hazard free. The safety related steps that can be taken during the product operation phase include:

1. Perform maintenance, operating, and emergency procedures audit.

2. Review operational equipment modifications and changes with respect to safety.

3. Analyze accidents and incidents, and initiate appropriate corrective measures.

9.4 SAFEGUARD DESIGN

Generally, in a situation where a designer is unable to achieve intrinsic safety, safeguards are used. In selecting safeguard material, the designer should consider such factors as: strength, stiffness, and durability [11]. Safeguards may be classified into such categories as [11]:

1. Interlocked guards.
2. Automatic guards.
3. Trip devices.
4. Adjustable guards.
5. Fixed guards.
6. Self-adjusting guards.

The selection of an interlock guard depends on several factors: the risk of danger, the momentum of the machine, the consequences of machine or safety device failure, and the simplicity of the chosen system. Other factors involved with interlocked guards are as follows:

1. The machine cannot begin operation prior to closing the guard.
2. The guard cannot be removed or opened until the hazardous components come to a standstill.
3. The guards can either be sliding, removable, or hinged, and the interlocking mechanisms must be reliable and safe.

Automatic guards force a person's body or limbs out of a potential trap area in the event of a trap. Usually, automatic guards are restricted to cutting or chopping machines and presses (slow-moving, long-stroke) and require careful human factor considerations.

Trip devices ensure that if a hazardous part is approached beyond a safe limit, the part comes to a standstill and the machine reverses. It is important to ensure that the trip devices are turned on by the involuntary action of the person. The difficulties associated with trip devices include: the brake mechanism and badly adjusted brakes.

An adjustable guard is a fixed guard with an adjustable element. These guards are often used for toolroom and woodworking machines. However, the protection provided by this type of guard is normally poor.

A self-adjusting guard prevents admittance to the workpiece, except when the guard is forced open to allow the transfer of work elements.

R. T. Booth [11] proposed a list of questions intended to determine whether the safeguards are appropriate for the products. These questions include:

1. Is the safeguard easy to use?

2. How easy is it to override or misuse the safeguard?

3. Do the safeguard parts satisfactorily meet reliability and fail-safe requirements?

4. Is the safeguard easy to maintain in the field?

5. Does the safeguard handle expected machine/product/equipment failures effectively?

6. Are the instructions for the machine and the safeguards effective with respect to possible hazards?

7. Does the safeguard absolutely stop an approach leading to a hazard when operating normally in its correct position?

9.5 SAFETY ANALYSIS TECHNIQUES AND DESIGN GUIDELINES

The two techniques, failure modes and effect analysis (FMEA) and fault trees, which are used for design-stage reliability analyses can also be applied to safety analyses. Both techniques are briefly discussed in the following paragraphs. A more detailed discussion may be found in the chapter on reliability.

1. **Failure modes and effect analysis (FMEA).** This method was originally developed for use in the design and development of flight control systems [2]. The method can also be used to evaluate design at the initial stage from the point of view of safety. Basically, the technique calls for listing the potential failure modes of each part, as well as the effects on the listed parts and on humans. The technique may be broken down into seven steps:

 a. Defining system boundaries and requirements.

 b. Listing all items (components/subsystems/etc.).

 c. Identifying each component and its associated failure modes.

 d. Assigning an occurrence probability or failure rate to each failure mode.

 e. Listing the effects of each failure mode on concerned items and people.

 f. Entering remarks for each possible failure mode.

 g. Reviewing and initiating appropriate corrective measures.

2. **Fault trees.** This technique was originally developed for reliability analyses of the Minuteman launch control system. It has also been applied successfully to safety analyses. The technique uses various symbols. It starts by identifying an undesirable event, called the top event, and then successively asking the question, "How could this event occur?" This process continues until the fault events need not be developed further. If occurrence data are known for the basic or primary fault event, the occurence measure for the top event can be calculated.

A safety analyst will evaluate an engineering design from various perspectives. Useful guidelines for this evaluation include the following:

1. Ensure that the design is safe with respect to gravity and balance.

2. Ensure the safety of design and operation procedures with respect to heavy part and high-energy spring maintenance.

3. Ensure that the incorrect assembly of parts, which may cause accidents, is prevented.

4. Ensure that for the maintenance of electric and high-pressure fluid services, the operating and design procedures are safe.

5. Ensure that the design enhances the safety of periodic inspections of critical parts by providing adequate accessibility and visibility.

9.6 PRODUCT SAFETY PROGRAM MANAGEMENT

The managerial aspects of product safety are as important as the safety-related tasks themselves. Managerial related safety tasks include: developing the company policy on product safety, assigning responsibilities for effective implementation of that policy, ensuring the availability of adequate funds for the safety program, developing safety training programs, and conducting periodic safety audits [10]. The primary purpose of these audits is to observe the performance of the product safety program. Therefore, the audits are used to determine the degree of compliance with the company safety policy, the effects of trade-offs on the safety of the product, the effectiveness of the coordination and integration of safety related activities, the status of safety functions within each designated area, etc.

9.6.1 PRODUCT SAFETY GROUPS

The safety organization within a company is only one of the groups that contribute to product safety. The other groups, depending on company size, product manufactured, etc., may include the following:

1. Manufacturing.

2. Purchasing.

3. Reliability.

4. Quality assurance.

5. Test engineering.

6. Legal services.

7. Customer relations.

8. Technical publications.

9. Field services.

10. Marketing.
11. Design.

In particular, the design engineering group in conjunction with the safety professionals, performs various safety related tasks: evaluating designs from the safety aspect and initiating appropriate corrective measures; evaluating the effectiveness of safety criteria developed by the safety specialist; identifying safety devices that need to be used in the product; highlighting the means to verify the expected performance of safety devices used; and ensuring the incorporation of pertinent safety information in manufacturing drawings.

9.6.2 PRODUCT SAFETY ENGINEER

Usually, in a fair sized design project, the product safety engineer is responsible for the safety aspects. The appointment of the product safety engineer must take into consideration the tasks involved and the candidate's qualifications. Both factors must be compatible for the safety engineer to be effective. Some of the tasks performed by a product safety engineer include:

1. Developing safety program plans.
2. Performing safety analyses.
3. Participating in design reviews.
4. Supervising subcontractors' safety efforts.
5. Collecting appropriate safety related information.
6. Providing advice on safety matters.
7. Interacting with outside safety agencies.
8. Developing innovative safety approaches.
9. Providing appropriate training.

9.6.3 PRODUCT SAFETY COSTS

Just as the overall product life-cycle cost, the product safety life-cycle cost can be estimated. The life-cycle cost associated with product safety may be obtained from the following relationship:

$$SLCC = SNRC + SRC \qquad \textbf{[9.1]}$$

where

$SLCC$ is the life-cycle cost associated with product safety.

$SNRC$ is the nonrecurring product safety cost (i.e., cost during product design and manufacturing phases).

SRC is the recurring product safety cost (i.e, cost during product operation phase).

Examples of nonrecurring product safety cost elements include: safety analyses, safety devices, and safety related documentation. Some of the recurring costs include: maintenance costs, insurance premiums (if applicable), accident cost, etc. The cost of an accident, to the manufacturer of a product, may be estimated from the following relationship [12]:

$$C_{ma} = C_p + C_{ai} + C_i + C_{pc} + C_{is} + C_f + C_{pd} + C_m \qquad \textbf{[9.2]}$$

where

C_{ma} is the cost of an accident to a product manufacturer.

C_p is the cost of preventive actions to eliminate recurrences.

C_{ai} is the cost of an accident investigation.

C_i is the cost of the increased insurance premium.

C_{pc} is the cost associated with the loss of public confidence.

C_{is} is the cost associated with the settlement of death or injury claims.

C_f is the cost associated with legal actions.

C_{pd} is the cost associated with property damage not covered by insurance.

C_m is the cost of miscellaneous activities.

9.7 TYPICAL SAFETY-RELATED HUMAN BEHAVIOR

Human behavior plays an important role in product safety. During the product design stage, a knowledge of human behavior could lead to improved safety measures in the design. Typical human behavior that may result in injury from an unsafe act include the following [3, 13, 14]:

1. People frequently fail to read, or incorrectly read, and/or understand instructions, labels, and scale markers.

2. People under emergency conditions often act irrationally.

3. People are generally not very good at estimating distances, velocities, or clearances. They tend to underestimate large distances and overestimate short distances.

4. In general, people use their hands to test, examine, or explore.

5. People fail to look where they put their hands and feet, particularly in surroundings that are familiar to them.

6. Usually, people are reluctant to admit judgmental or perceptual mistakes or errors.

7. People generally underestimate weight if it is compact and overestimate it if it is bulky.

8. People generally underestimate speed if an object is decelerating and overestimate it if the object is accelerating.

9. People generally underestimate cold temperatures and overestimate hot temperatures.

10. People generally underestimate the "unpleasant event" occurrence probability and overestimate the "pleasant event" occurrence probability.

11. People generally underestimate height when looking up and overestimate it when looking down.

12. People often underestimate horizontal distance.

13. People usually fail to recheck maintenance or operating procedures for possible omissions or errors.

14. Only a small percentage of people recognize that they are unable to see effectively, either due to poor illumination or poor eyesight.

15. Most people do not take the proper time to observe safety precautions, or read labels or instructions.

16. Most people continue to use faulty items, even when they suspect these items could be hazardous.

W. Haddon [15] proposed the following strategies to minimize safety-related problems:

1. Avert hazard formation.
2. Eliminate existing hazards.
3. Reduce hazard magnitude
4. Change the primary qualities of hazards.
5. Counter the damage caused by hazards.
6. Make items hazard damage resistant.
7. Change the hazard distribution rate.

9.8 SELECTED SAFETY INFORMATION SOURCES

This section presents selected safety information sources, including: books and related documents, journals, data sources, and organizations [5]. More comprehensive source information is given in References 5 and 6.

9.8.1 BOOKS AND RELATED DOCUMENTS

1. Hammer, W. *Product Safety Management and Engineering*. Englewood Cliffs, NJ: Prentice-Hall, 1980.

2. Brown, D.B. *Systems Analysis and Design for Safety*. Englewood Cliffs, NJ: Prentice-Hall, 1976.

3. Raouf, A.; and B.S. Dhillon. *Safety Assessment: A Quantitative Approach.* Boca Raton, FL: Lewis Publishers, 1994.

4. Gloss, D.S.; and M.G. Wardle. *Introduction to Safety Engineering.* New York: Wiley, 1984.

5. Dhillon, B.S. *Robot Reliability and Safety.* New York: Springer Verlag, 1991.

6. Wells, G.L. *Safety in Process Plant Design.* New York: Wiley, 1980.

7. *Design Handbook-System Safety.* AFSC DH1-6 AFSC. Washington, D.C.: U.S. Department of Defense, July 1967.

8. *Systems Safety Program for System and Associated Subsystem and Equipment-Requirements.* MIL-STD-882. Washington, D.C.: Department of Defense, July 1969.

9. Kolb, J.; and S.S. Ross. *Product Safety and Liability.* New York: McGraw-Hill, 1980.

10. Dhillon, B.S. *Reliability Engineering in System Design and Operation.* New York: Van Nostrand Reinhold, 1983, Chapter 7. (This chapter also lists over 130 references related to safety.)

9.8.2 JOURNALS

1. *Product Safety News.* Durham, NC: Institute for Product Safety.

2. *Hazard Prevention.* Seabrook, TX: System Safety Society.

3. *Professional Safety.* Park Ridge, IL: American Society of Safety Engineers.

4. *Journal of Safety Research.* Chicago, IL: National Safety Council.

5. *Accident Analysis and Prevention.* Elmsford, NY: Pergamon Press.

6. *Occupational Accidents Journal.* Amsterdam, The Netherlands: Elsevier Scientific Publishing.

9.8.3 DATA SOURCES

1. Safety Research Information Service, National Safety Council, 444 North Michigan Avenue, Chicago, IL.

2. Nuclear Safety Information Center, Oak Ridge National Laboratories, P.O. Box Y, Oak Ridge, TN.

3. Computer Accident/Incident Reporting System, System Safety Development Center, EG & G, P.O. Box 1625, Idaho Falls, ID.

4. Defense Technical Information Center, Defense Logistics Agency, Cameron Station, Alexandria, VA.

5. Safety Science Abstract Journal, Cambridge Scientific Abstract, Inc., 5161 River Road, Washington, D.C.

6. Government Industry Data Exchange Program (GIDEP), GIDEP Operations Center, Navy Fleet Analysis Center, Corona, CA.

7. Hazardline, Occupational Health Services, 515 Madison Ave., Dept. D, New York, NY.

8. NIOSHTIC, National Institute for Occupational Safety and Health, Parklawn Building, 5800 Fisher Lane, Rockville, MD.

9. Loss Management Information System, Gulf Canada Ltd., 800 Bay Street, Toronto, Ontario.

9.8.4 ORGANIZATIONS

1. American Society of Safety Engineers, 850 Busse Highway, Park Ridge, IL.

2. System Safety Society, 14252 Culver Dr., Suite A-261, Irvine, CA.

3. National Safety Council, 444 North Michigan Ave., Chicago, IL.

4. Industrial Accident Prevention Association of Ontario, 2 Bloor Street East, Toronto, Ontario.

5. Academy of Product Safety Management, 5010 A Nicholson Lane, Rockville, MD.

9.9 CASE STUDY: BAY AREA RAPID TRANSIT SYSTEM (BART)

This $1 billion system for the San Francisco bay area was put into operation in the early 1970s [16]. The system originally encompassed five counties: San Francisco, San Mateo, Marin, Contra Costa, and Alameda.

Some of the performance specifications for BART vehicles [17] on level tangent track were: (a) maximum controllable running speed of at least 113 km/h; (b) maximum acceleration (instantaneous rate) of 5.6 km/h/s; and (c) maximum deceleration (instantaneous rate) of 5.6 km/h/s.

On October 2, 1972, a southbound BART train overshot the Fremont terminal, plunging the lead car onto a sand embankment, as shown in Figure 9.1 [18]. At the time of the incident, the BART train was operating under the complex automatic (computer-controlled) train-control (ATC) system and the train's operator only managed to override that system manually just prior to impact. Fortunately, this incident resulted in no fatalities or serious injuries. According to a BART official, the failure of a crystal oscillator in the ATC was responsible for this mishap.

On December 19, 1972, the California State Senate Public Utilities and Corporations Committee appointed a three-person panel to investigate the alleged deficiencies in the BART system. On January 31, 1973, the panel concluded that among other things, the current BART restricted mode of operation was reasonably safe, and was an appropriate interim action.

Figure 9.1   BART train incident in October 1972

Lonnie Wilson/Oakland Tribune

In addition, the panel concluded that:

1. The current BART ATC system design was inadequate to provide satisfactory passenger safety for full-scale operation.
2. The present design could be modified to provide satisfactory passenger safety for all service conditions.

9.10 PROBLEMS

1. Write a brief essay on the history of safety considerations.
2. Discuss system safety functions.
3. Discuss safety-related tasks for the four product life-cycle phases: concept, design and development, manufacture, and operation.
4. Discuss the following safety devices:
   a. Trip devices.
   b. Fixed guards.
   c. Interlocked guards.
5. What groups in a large organization contribute to product safety?
6. What are the attributes of a product safety engineer?
7. What are the product safety responsibilities of a design engineer?
8. List at least 10 typical human behaviors.

9. What are the functions of a system safety engineer?

10. Discuss at least three military documents concerned with system safety.

REFERENCES

1. Hammer, W. *Product Safety Management and Engineering.* Englewood Cliffs, NJ: Prentice-Hall, 1980.

2. *Accident Facts.* Chicago: National Safety Council, 1981.

3. Gloss, D.S.; and M.G. Wardle. *Introduction to Safety Engineering.* New York: Wiley, 1984.

4. Heinrich, H.W. *Industrial Accident Prevention.* 4th ed. New York: McGraw-Hill, 1959.

5. Raouf, A.; and B.S. Dhillon. *Safety Assessment: A Quantitative Approach.* Boca Raton, FL: Lewis Publishers, 1994.

6. Ferry, T.S. *Safety Program Administration for Engineers and Managers.* Springfield, IL: Charles C. Thomas Publishers, 1984.

7. Dhillon, B.S. *Reliability Engineering in Systems Design and Operation.* New York: Van Nostrand Reinhold, 1983.

8. *Accident Prevention Manual.* 9th ed. Chicago: National Safety Council, 1986.

9. Rodgers, W.P. *Introduction to System Safety Engineering.* New York: Wiley, 1971.

10. Hammer, W. *Product Safety Management and Engineering.* Englewood Cliffs, NJ: Prentice-Hall 1980.

11. Booth, R.T. "Industrial Design and Safety." In *Industrial Design in Engineering*, ed. C.H. Flurscheim. Berlin: Springer-Verlag, 1983, pp. 245–262.

12. Dhillon, B.S. *Robot Reliability and Safety.* New York: Springer-Verlag, 1991.

13. Nertney, R.J.; and M.G. Bullock. *Human Factors in Design.* Report No. ERDA-76-45-2. Washington, D.C.: U.S. Department of Energy, Energy Research and Development Administration, 1976.

14. Salvendy, G. "Human Factors in Planning Robotic Systems." In *Handbook of Industrial Robotics,* ed. S.Y. Nof. New York: Wiley, 1985, pp. 639–664.

15. Haddon, W. "The Basic Strategies for Reducing Damage from Hazards of All Kinds." *Hazard Prevention*, September/October 1980.

16. Friedlander, G.D. "The BART Chronicle" *IEEE Spectrum,* September 1972, pp. 34–46.

17. Friedlander, G.D. "BART's Hardware: from bolts to Computers." *IEEE Spectrum,* October 1972, pp. 60–72.

18. Friedlander, G.D. "Bigger Bugs in BART?" *IEEE Spectrum,* March 1973, pp. 32–37.

chapter
10

HUMAN FACTORS CONSIDERATIONS

10.1 INTRODUCTION

Human factors may be described as the body of knowledge related to human abilities, limitations, etc. In 1898 Frederick W. Taylor, who was probably the first human factors engineer, studied how to design effective shovels [1]. Then, in 1911, Frank B. Gilbreth studied bricklaying and invented the scaffold, which increased the rate of bricklaying to 120–350 per man–hour [2]. In 1918, the United States Air Force (USAF) played a key role in the development of the field by establishing human factors research laboratories at the Wright-Patterson Air Force Base and the Brooks Air Force Base [3]. At the end of World War II, human factors engineering was a recognized discipline, and today, there are several research journals, conference, and textbook publications devoted exclusively to human factors.

The design of engineering systems must take into consideration the impact of human factors. This chapter discusses those aspects of the human factors discipline that directly or indirectly affect engineering design.

10.2 HUMANS AND MACHINES

Human behavior is critical to the success of an engineering system. Therefore, during the design phase, it is important to take typical human behaviors into consideration. Examples of typical human behaviors monitored over the years are as follows [4]:

1. Humans are often reluctant to admit errors.
2. Humans usually perform tasks while thinking about other things.
3. People frequently misread or overlook instructions and labels.
4. People often respond irrationally in emergency situations.
5. A significant percentage of people become complacent after successfully handling dangerous items over a long period of time.
6. Most people fail to recheck outlined procedures for errors.
7. People are generally poor estimators of speed, clearance, or distance. Interestingly, they frequently overestimate short distances and underestimate large distances.

8. People are generally too impatient to take the time needed to observe precautions.

9. People are often reluctant to admit that they cannot see objects well enough, due either to poor eyesight or to inadequate illumination.

10. People generally use their hands for examining or testing.

In designing engineering systems, the engineers must decide whether a task is to be assigned to a human or a machine. To make such decisions effectively, the designer must first have a knowledge of human and machine characteristics. Important comparable machine and human characteristics [5] are as follows:

Machine   Some of these characteristics may be more applicable to robots than to general machines.

1. Machines require no motivation.
2. Machines are consistent, unless there are failures.
3. Machines require periodic maintenance, but are free from fatigue.
4. Machines possess only limited intelligence and judgmental capability.
5. Machines possess a poor inductive capability, but a good deductive ability.
6. Machines are prone to failures.
7. Machines have a fairly fast reaction time to external signals.
8. Machines are not easily affected by the environment; therefore they are particularly useful in applications in unfriendly environments.
9. The machine memory is not affected by absolute and elapsed times.

Human

1. Humans require some degree of motivation.
2. Human consistency can be low.
3. Humans are subject to fatigue, which increases with the number of hours worked and diminishes with rest.
4. Humans possess a high degree of intelligence, and are capable of applying judgment to solve unexpected problems.
5. Humans have inductive capabilities.
6. Humans may be absent from work, due to such factors as personal matters, illness, training, and strikes.
7. Human reaction time is slow compared to that of machines.
8. Humans are affected by environmental factors, such as temperature, noise, and hazardous materials, and they need air to breathe.
9. Human memory may be constrained by elapsed time, but it has no capacity limitations.

10.3   HUMAN SENSORY CAPACITIES AND VISUAL AND AUDITORY MESSAGES

Human sensory capacities [6,7] include; sight, noise, touch, motion, and vibration. These are described in the following paragraphs.

10.3.1   SIGHT

Human eyes "see" differently from different angles or positions. For example, looking straight ahead, the eyes can perceive all colors. However, with an increase in the viewii.g angle, human perception begins to decrease. For vertical and horizontal situations, the limits of color vision are as given in Table 10.1.

Table 10.1   Limits of Color Vision

Situation	Color and Degrees					
	Green	Blue	Yellow	White	Green–Red	Red
	n		w			
Vertical	40°	80°	95°	130°	—	45°
Horizontal	—	100°	120°	180°	60°	—

Also at night or in the dark, small–sized orange, blue, green, and yellow warning lights are impossible to distinguish from a distance. According to various studies, they all appear to be white [6,7].

10.3.2   NOISE

The performance quality of a task requiring intense concentration may be affected by noise. It is an established fact that noise contributes to peoples' feelings such as irritability, boredom, and well-being. A noise level below 90 decibels (dB) is considered harmless, and above 100 dB is unsafe. Noise levels in excess of 130 dB are considered unpleasant and may actually be painful. Furthermore, above-normal noise may make verbal communication (such as between operators and maintainers) impossible, and may even damage their hearing [6,7].

10.3.3   TOUCH

Touch adds to, or may even replace, the information transmitted to the brain by the eyes and ears. For example, it is possible to distinguish control knob shapes by touch alone, which could be valuable if, say, power goes out and there is no light. This vital sensory capacity must especially be taken into consideration when the design professional must depend completely on touch controls [6,7].

10.3.4 VIBRATION AND MOTION

It is now an accepted fact that the poor performance of physical and mental tasks by equipment operators and maintainers could be due partially or fully to vibrations. For example, eye strain, headaches, and motion sickness could result from low-frequency, large-amplitude vibrations. Useful guidelines for reducing the effects of vibration and motion are as follows [6,7]:

1. Elimate vibrations with an amplitude greater than 0.08 millimeter.
2. Use such devices as shock absorbers and springs wherever possible.
3. Use damping materials or cushioned seats to reduce vibrations wherever possible.
4. Since seated humans are most affected by vertical vibrations, use this information to develop reduced vertical vibrations.
5. The resonant frequency of the human vertical trunk, in the seated position, is between 3 and 4 cycles per second. Therefore, avoid any seating design that would result in or would transmit a 3 to 4 cycles per second vibration.

10.3.5 VISUAL AND AUDITORY MESSAGES

During equipment design, some decisions may involve whether to use visual or auditory channels to communicate messages. The situations in which the messages would be sent must be carefully examined. The situations under which either a visual or an auditory channel should be considered are as follows [8]:

Visual Channel

1. Long message.
2. Complex message.
3. Receiving end has above-normal noise level.
4. Received information does not call for a fast response.
5. Person receiving the information is in a fixed position.
6. Transmitted message will be referred to at a later time.
7. Auditory system of person receiving the message is or would be overburdened.

Auditory Channel

1. Short message.
2. Simple message.

3. Receiving end is too bright.

4. Visual system of person receiving the message is or would be overburdened.

5. Person receiving the information is mobile.

6. Transmitted message will not be referred to at a later time.

10.4 WARNINGS, AND HUMAN FACTORS CHECKLISTS

There are basically three design approaches that can be used to make a product safe for use [9]: eliminate unsafe features, provide protection against possible hazards, and provide necessary warnings against possible misuse.

Warnings are relatively inexpensive to produce, but their effectiveness is not guaranteed, since they may be totally ignored. Nevertheless, it is still essential to pay particular attention to the wording and display of warnings. Warnings should correspond to the potential dangers and should distinguish between the following [10]:

1. **Caution.**   This is to be used under conditions where hazards or unsafe practices *may* lead to minor personal injury, and/or minor damage to the product or to property.

2. **Danger**.   This is to be used where immediate hazards (if they occur) *would* lead to severe personal injury or death.

3. **Warning**.   j193

This is to be used where hazards or unsafe acts (if they occur) *could* lead to severe personal injury or death.

In incorporating human factors into the designs of engineering systems, various designers have developed a checklist of questions to be addressed. These questions can be specifically tailored to each situation. Examples of the questions are as follows:

1. Can each control device be easily identified?

2. Is the visual display arrangement optimized?

3. What sensory channel would be most suitable for messages to be communicated through the displays?

4. Does the design effectively use human decision making and adaptive capabilities?

5. Are the controls effectively designed with respect to shape, size, accessibility, etc.?

6. Does the design effectively address the possibility of grouping the tasks to be performed into jobs?

7. Does the workspace design take human factor principles into consideration?

8. Are control devices compatible with their corresponding displays, with respect to human factors?

9. For satisfactory levels of human performance, were such environmental factors as noise, temperature, and illumination taken into consideration?

10. Was sufficient attention paid to complementing work aids and training?

10.5   HUMAN ERROR AND RELIABILITY

Human error is the failure to perform a given task, or the performance of a prohibited action resulting in a disruption of scheduled operations, or in damage to property and equipment or even in personal injury. In contrast, human reliability is the probability of a given task at any stage in system operation being performed, within a given minimum time limit if applicable [2].

In equipment design, wherever human action is involved, the potential for human error must be carefully considered. According to one study [11], from 20 to 50 percent of all equipment failures could be attributed to humans. The reasons for human errors include:

1. Poor skill/training of personnel, such as operators, maintainers, and production workers.
2. Ineffective product design.
3. Improper or poor-quality tools.
4. Poor lighting in the work area.
5. Poor-quality maintenance and operating procedures.
6. Poor surrounding environments.
7. Task complexity.

10.5.1   HUMAN ERROR CONSEQUENCES

The consequences of human error vary from one situation to another, and the severity of those consequences may range from minor to critical. With respect to equipment, possible consequences may be categorized as:

1. Operation is prevented.
2. Operation is significantly delayed.
3. Operation is insignificantly delayed.

10.5.2   HUMAN ERROR CATEGORIES

In general, human error can be classified into seven categories, as follows [11, 12]:

1. **Operation error.**   This type of error is attributed to operating personnel. The possible causes include: improper procedures, poor surrounding environment, task complexity, overload conditions, operator carelessness, inadequate personnel training or selection, and incorrect operating procedures.

2. **Maintenance error.**   These errors occur in the field and are attributable to maintenance personnel. Two examples are: the use of the wrong grease to lubricate the equipment, and the incorrect calibration of equipment. According to D. Meister

[13], the probability of a maintenance error occurring may increase as the equipment gets older, because the maintenance frequency usually increases.

3. **Design error.**   These errors reflect inadequate designs. One example of a design error is placing the controls and the displays so far apart that the operator has difficulty using them effectively. The basic categories of design errors are: assigning an incorrect task to a human, failing to implement human requirements in the design, and neglecting to ensure the effectiveness of the human and machine interactions. The various causes of design errors include: inadequate analysis of the system requirements, designer's bias toward a specific design, insufficient time spent on the design effort, etc.

4. **Inspection error.**   These errors are committed by inspection manpower. One example of this type of error is the acceptance of an out-of-tolerance item or the rejection of an in-tolerance item. The purpose of inspection is to uncover all items with defects. Unfortunately, it is generally not practical to achieve 100 percent inspection accuracy. In fact, according to R. L. McCormack [14], inspection effectiveness averages somewhere in the neighborhood of 85 percent.

5. **Fabrication error.**   Fabrication errors result from poor workmanship during product assembly. Typical examples of such errors are: using the wrong soldering materials and parts, wiring a part incorrectly, and assembling a system incompatibly with blueprints. Causes of production errors include: poor blueprints, inadequate illumination, poor work layout design and excessive noise.

6. **Installation error.**   These errors are attributable to incorrect or incomplete installation of a product. A prime cause of such errors is the failure of humans to follow the instructions or blueprints.

7. **Handling error.**   Handling errors are due to the inappropriate or improper storage or transport of a product, and can result in damage to the product. Improper packaging for shipment is an example of such errors.

8. **Contributory error.**   This classification covers those errors that are difficult to identify as either human or hardware.

10.5.3   HUMAN ERROR PREVENTION METHODS

Several approaches have been developed to prevent the occurrence of human error [2]. Three of these are:

1. **Man–machine systems analysis method.**   This method reduces unwanted effects caused by human error to an acceptable level. The method is comprised of 10 steps.

2. **Error cause removal program.** This method emphasizes preventive measures rather than remedial ones and is useful for reducing human error to a tolerable level during production operations. The method is based on production worker participation.

3. **Quality circle method.** This method was originated in Japan in 1963 to solve quality-related problems. The method is similar to the error cause removal program. Some of their elements are much the same: participative democracy concept, problem solving inclination, and crossover among management levels.

For more information on these three methods, consult Reference 2.

10.5.4   CASE STUDY: THREE MILE ISLAND NUCLEAR POWER PLANT ACCIDENT

A nuclear plant, shown in Figure 10.1, is located on Three Mile Island, ten miles southeast of Harrisburg, Pennsylvania. On March 28, 1979, an accident occurred at the number-two reactor. The accident is regarded as a crucial milestone in the history of the peaceful use of atomic energy throughout the world. President Jimmy Carter established a commission to investigate the accident, under the Chairmanship of John G. Kemeny, president of Dartmouth College [15]. The Commission reported that the basic cause of the reactor mishap was an interwoven chain of mechanical, human, and institutional malfunctions. The report of the commission included 44 recommendations to President Carter. Two of the recommendations were [15, 16]:

Figure 10.1   Three Mile Island nuclear power plant.

- Improve the warning display panels in the control rooms such that the attending operators can respond more efficiently to emergency situations.

- Provide monitoring and recording equipment for continuous recording of all crucial plant measurements and conditions.

According to a report published in *New Scientist* [15], the Commission still failed to provide clear answers to some questions, such as why the emergency feedwater pump valves remained closed for eight minutes after the beginning of the accident.

10.6 HUMAN FACTORS DATA

Human factors data are important elements in the analyses of engineering designs. The various types of human factors data include: human error occurrence rates, body dimensions and weights, permissible noise exposure per unit time, energy expenditure per grade of work, and so on. The data may exist in a number of forms such as:

1. Mathematical functions and expressions.
2. Expert judgments.
3. Experience, and common sense.
4. Design standards.
5. Established principles.
6. Graphic representations.
7. Quantitative data tables.

10.6.1 DATA SOURCES

There are several ways of collecting human factors data. Some of these are listed here [17].

1. **Previous experience.**   These data are obtained from similar cases that have occurred in the past.

2. **Published literature.**   This includes books, journals, conference proceedings, and reports.

3. **Product development phase.**   This is a good source to collect various types of human factors related data.

4. **Published standards.**   These documents are published by government agencies, professional societies, etc.

5. **Test reports.** These are the results from tests conducted on the manufactured goods.

6. **User experience reports.** These are the reports that reflect the users' experiences with the equipment.

10.6.2 DATA DOCUMENTS

There are many useful resources from which human factors data can be retrieved. Some of those are as follows:

1. *Human Engineering Guide to Equipment Design.* ed. H.P. Van Cott and R.G. Kinkade. New York: Wiley, 1972.
2. Woodson, W.E. *Human Factors Design Handbook.* New York: McGraw-Hill, 1981.
3. McCormick, E.J.; and M.S. Sanders. *Human Factors in Engineering and Design.* New York: McGraw-Hill, 1982.
4. "Anthropometry for Designers." *Anthropometric Source Book* 1, 1024. Houston, TX: (1978), National Aeronautics and Space Administration.
5. Dhillon, B.S. *Human Reliability with Human Factors.* New York: Pergamon Press, 1986.
6. *Handbook of Human Factors.* ed. G. Salvendy. New York: Wiley, 1987.
7. Swain, A.; and H. Guttmann. *Handbook of Human Reliability Analysis with Emphasis on Nuclear Power Plant Applications.* Washington, D.C.: U.S. Nuclear Regulatory Commission, 1980.
8. *Bioastronautics Data Book.* ed. J. F. Parker and V. R. West, NASA-SP-3006. Washington, D.C.: U.S. Government Printing Office.
9. White, R.M. "The Anthropometry of United States Army Men and Women: 1946–1977." *Human Factors* 21 (1979), pp. 473–482.
10. *Ethnic Variables in Human Factors Engineering.* ed. A. Chapanis. Baltimore: Johns Hopkins University Press, 1975.
11. Meister, D.; and D. Sullivan. *Guide to Human Engineering Design for Visual Displays.* no. AD 693237 (1969), Springfield, VA: National Technical Information Service (NTIS).
12. *Data Base (0025), U.S. Army.* Alexandria, VA: PERI-II, Army Research Inst., 1975.
13. *APS Data.* Wayne, PA; Applied Psychological Services (APS), 1987.
14. *Human Engineering Design Criteria for Military Systems.* MIL-H1472D. U.S. Department of Defense, Philadelphia, PA: 1972.
15. *Lighting Handbook.* New York: Illumination Engineering Society, 1971.

16. Tillman, P.; and B. Tillman. *Human Factors Essentials.* New York: McGraw-Hill, 1991. (This book lists several additional sources of human factors information.)

10.6.3 SELECTED QUANTITATIVE DATA

Human factors professionals have accumulated several types of data, which are available in varying forms. Tables 10.2–10.5 present selected data on noise exposure (guidelines by U.S. Occupational Safety and Health Administration), structural body dimensions, energy expenditure per grade of work, and human error rates.

Table 10.2 **S**elected acceptable levels of noise exposure

Exposure time in hours (per day)	Sound levels in decibels (dBA)
1	105
2	100
4	95
8	90
0.5	110

Table 10.3 **S**elected body dimensions

Body Feature	Male Body Dimensions in Inches	
	5th Percentile	95th Percentile
Normal sitting height	31.6	36.6
Seat breadth	12.2	15.9
Height	63.6	72.8
Knee height	19.3	23.4

Table 10.4 **E**nergy expenditure per grade of work

Work Type	Energy Expenditure in Kcal. per Minute
Light	2.5–5.0
Heavy	7.5–10
Moderate	5.0–7.5

Table 10.5 Selected data for human error rates

Task Description	Human Error Rate per Million Operations
Tightening nut and bolt	4,800
Reading gauge	5,000
Reading instructions	64,500
Installing nut and bolt	600
Soldering connectors	6,460

10.7 SELECTED HUMAN FACTORS FORMULAS

This section presents several human factors formulas that are useful in designing engineering systems.

10.7.1 CHARACTER HEIGHT

To determine the character height for the labels on panel fronts, Adams and Peters [17] propose the following formula:

$$H_c = CF_c + CF_{vi} + (0.0022)VD \qquad \text{[10.1]}$$

where

CF_c is the correction factor for criticality. For critical and noncritical markings, respectively, the values are 0.075 and 0.

H_c is the character height.

CF_{vi} is the correction factor for the viewing and illumination conditions. For specified values, see Reference [18].

VD is the viewing distance, expressed in inches.

10.7.2 LIFTING LOAD

The formula for estimating the maximum lifting load for a person is [19]:

$$ML = k(MS) \qquad \text{[10.2]}$$

where

ML is the maximum lifting load.

MS is the isometric back muscle strength.

k is a constant, whose values are 1.1 and 0.95, respectively, for males and females.

10.7.3 REQUIRED REST PERIOD

For many activities, rest periods are incorporated into the daily work schedule. K. F. H. Murrell [20] presents the following formula for estimating the required duration of a rest period:

$$RT = WT(AE - EE)/(AE - RL) \qquad [10.3]$$

where

RT is the rest period, expressed in minutes.

WT is the total working time, expressed in minutes.

AE is the average energy expenditure per minute of work, expressed in kilocalories (Kcal/min).

EE is the standard energy expenditure, expressed in kilocalories per minute. In situations when no data are available, its value may be taken as 5 Kcal/minute.

RL is the resting level. Its approximate value is 1.5 Kcal/minute.

10.7.4 HUMAN RELIABILITY

Many day-to-day tasks performed by humans are time-continuous tasks: monitoring, a scope, driving motor vehicles, aircraft maneuvering, and so on. The reliability with which humans will perform such tasks correctly may be obtained from the following relationship [2]:

$$HR(t) = e^{\left[-\int_0^t he(t)dt\right]} \qquad [10.4]$$

where

$HR(t)$ is the human reliability at time t.

$he(t)$ is the time-dependent human error rate.

Example 10.1 Assume that a person is performing a time-continuous task and the associated human error rate is 0.001/hour. Calculate the reliability of performing the task correctly for a 4-hour time period.

Solution

Substituting the given data into Equation 10.4 yields

$$HR(4) = e^{[-(0.001)(4)]}$$
$$= 99.6\%$$

This result means there is a 99.6 percent chance that the person will perform the task correctly over the 4-hour period.

10.7.5 GLARE CONSTANT

Glare can be a serious problem in the work environment. D. J. Oborne [21] presents the following formula for estimating the value of the glare constant:

$$GC = (SA)^{0.8}(SL)^{1.6}/(LGB)(ASD)^2 \qquad \textbf{[10.5]}$$

where

GC is the value of the glare constant.

SA is the solid angle subtended at the eye by the source.

ASD is the angle between the viewing direction and the glare source direction.

SL is the luminance of the source.

LGB is the general background luminance.

10.8 CASE STUDY: LARGE STACKER FORKLIFT TRUCK ACCIDENT

In Arizona, a large stacker forklift truck was involved in a fatal accident, due to a defective design of the control system display [22]. The mishap occurred during the transport of concrete sewer pipe sections. The truck operator intended to tilt the mast back to position the pipe section for unloading; however, the mast tilted forward and the pipe rolled off the forks. The subsequent investigation revealed that the real cause of the accident was the control toggle switch for mast tilt manipulation functions. The switch design was contrary to normal human behavior; that is, the switch had to be moved forward to tilt the mast backward and vice-versa. The operator unintentionally pulled the toggle switch backward, causing the fatal accident.

10.9   PROBLEMS

1. Compare the facts about automatic machines and humans.
2. Write an essay on the history of human factors.
3. Describe the following terms:
   a. Ergonomics.
   b. Human error.
   c. Human reliability.
   d. Human engineering.
4. What are the human sensors? Discuss each in detail.
5. Discuss the five most important categories of human error.
6. Discuss possible measures for preventing human error.
7. What are the possible sources for collecting human factor data?
8. After testing sets of data concerning the performance of a time continuous task, it is established that the times to human error are exponentially distributed, with the mean of 100 hours. Calculate the human reliability for a three-hour mission.

REFERENCES

1. Chapanis, A. *Man-Machine Engineering.* Belmont, CA: Wadsworth, 1965.
2. Dhillon, B.S. *Human Reliability with Human Factors.* New York: Pergamon Press, 1986.
3. Meister, D.; and G.F. Rabideau. *Human Factors Evaluation in System Development.* New York: Wiley, 1965.
4. Nertney, R.J.; and M.G. Bullock. *Human Factors in Design.* Report No. ERDA-76-45-2. Washington, DC: U.S. Department of Energy, Energy Research and Development Administration, 1976.
5. Dhillon, B.S. *Robot Reliability and Safety.* New York: Springer-Verlag, 1991.
6. *Engineering Design Handbook, Maintainability Guide for Design.* AMCP 706-134. Washington DC: U.S. Army Material Command, 1972.
7. Dhillon, B.S. *Quality Control, Reliability, and Engineering Design.* New York: Marcel Dekker, Inc., 1985.
8. Vidosic, J.P. *Elements of Design Engineering.* New York: The Ronald Press, 1969.
9. Fowler, F.D. "Failure to Warn: A Product Design Problem." *Proc. of the Symposium on Human Factors and Industrial Design in Consumer Products.* Medford, MA: Tufts Univ., 1980, pp.241–250.

10. McCormick, E.J.; and M.S. Sanders. *Human Factors in Engineering and Design.* New York: McGraw-Hill, 1982.

11. Meister, D. "The Problem of Human-Initiated Failures." *Proc. of the National Symposium on Reliability and Quality Control,* 1962, pp. 234–239.

12. Dhillon, B.S.; and C. Singh. *Engineering Reliability: New Techniques and Applications.* New York: Wiley, 1981.

13. Meister, D. *Human Factors: Theory and Practice.* New York: Wiley, 1976.

14. McCormack, R.L. *Inspector Accuracy: A Study of the Literature.* Report No. SCTM 53-61 (14). Albuquerque, NM: Sandia Corporation, 1961.

15. Torrey, L. "The Accident at Three Mile Island." *New Scientist,* November 8, 1979, pp. 424–428.

16. Bignell, V.; and J. Fortune. *Understanding Systems Failures.* Manchester, U.K: Manchester University Press, 1984.

17. Peters, G.A.; and B.B. Adams. "Three Criteria for Readable Panel Markings." *Product Engineering* 30 (1959), pp.55–57.

18. Huchingson, R.D. *New Horizons for Human Factors in Design.* New York: McGraw-Hill, 1981.

19. Poulsen, E.; and K. Jorgensen, "Back Muscle Strength, Lifting and Stooped Working Postures." *Applied Ergonomics* 2 (1971), pp. 133–137.

20. Murrell, K.F.H. *Human Performance in Industry.* New York: Reinhold, 1965.

21. Oborne, D. J. *Ergonomics at Work.* Chichester: Wiley, 1982.

22. Ross, B. "What Is a Design Defect?" In *Structural Failure, Product Liability and Technical Insurance*, ed. H. P. Rossmanith. New York: Elsevier Science Publishing Company, Inc., 1984, pp. 44–46.

chapter

11

ECONOMIC ANALYSIS AND COST ESTIMATION

11.1 INTRODUCTION

The history of economics goes back thousands of years. For example, the early human history reveals that the concept of interest existed in Babylon 2,000 years before the birth of Jesus Christ. In fact, in those days, interest was paid on borrowed commodities (e.g., grain) in the form of grain or some other means [1]. In modern times, engineers concerned with the product design must make a number of economic decisions, based on sound economic analyses. These analyses might be concerned with maintenance and operational costs, manufacturing cost, etc. Design engineers must therefore possess some knowledge of at least basic principles of the engineering economics. Many people have developed various types of mathematical models for these analyses, many of which are based on empirical studies. This chapter discusses various aspects of economic analysis and cost estimation.

11.2 SIMPLE AND COMPOUND INTEREST

Simple interest is earned on the original amount only. On the other hand, compound interest is earned on the original amount and on the previous accumulated interest. For example, assume an original amount of $1,000. At 10 percent simple interest per year, a flat $100 would be added to the original amount each year. At 10 percent compound interest per year, the first year would earn $100, the second year would earn $100 plus $10, etc. Mathematically, the amount at the end of the first period is

$$A_1 = A_0 + A_0 i \qquad \textbf{[11.1]}$$

where

A_0 is the amount (or principal) at the beginning of the first period.

i is the interest rate per period.

A_1 is the amount at the end of the first period (or the beginning of the second).

Similarly, the amount at the end of the nth period is

$$A_n = A_0(1 + i)^n \qquad \text{[11.2]}$$

where

A_n is the amount at the end of the nth period.

n is the number of interest periods, called conversion periods (normally taken as years).

The present value of the amount at the end of the nth period from Equation 11.2 is

$$PV = A_0 = \frac{A_n}{(1 + i)^n} \qquad \text{[11.3]}$$

where PV is the present value of the amount at the end of the nth period.

A company borrows $100,000 to design a mechanical system, at the interest rate of 6 percent per annum for a period of 10 years. Calculate the total amount of money spent on this loan at the end of the 10-year period. | **Example 11.1**

Solution

Substituting the specified values in Equation 11.2 yields

$$A_{10} = (100,000)(1 + 0.06)^{10}$$
$$= \$179,084.70$$

The total amount of money spent at the end of the 10-year period will be $179,084.70.

The expected salvage value of a water pump after its 15-year predicted useful life is $60,000. The estimated value of the interest rate is 8 percent per year. Calculate the present value of the pump salvage value. | **Example 11.2**

Solution

Inserting the given data into Equation 11.3 yields

$$PV = \frac{(60,000)}{(1 + 0.08)^{15}} = \$18,914.50$$

The present value of the pump salvage value is $18,914.50.

11.3 FUTURE VALUE: UNIFORM PERIODIC PAYMENTS

This case is concerned with the future value (i.e., compound amount) of equal payments made at the end of each of n conversion periods (usually years). It is assumed that these payments are invested at the compound interest rate i per annum. The final future value formula is developed as follows:

The first payment is made at the end of the first conversion period. From Equation 11.2, its future amount at the end of the second conversion period is

$$A_1 = AP + (AP)i \qquad \textbf{[11.4]}$$
$$= AP(1+i)$$

where AP is the payment made at the end of each conversion period.
At the end of the second period, the amount AP is added to A_1; thus,

$$TA_1 = AP(1+i) + AP \qquad \textbf{[11.5]}$$

where TA_1 is the total amount at the end of the second period. Similarly, the future amount at the end of the third period is

$$A_2 = TA_1 + (TA_1)i = TA_1(1+i) \qquad \textbf{[11.6]}$$
$$= [AP(1+i) + AP](1+i)$$
$$= AP(1+i)^2 + AP(1+i)$$

The amount A_2 is added to AP, and the total amount at the end of the third period is

$$TA_2 = A_2 + AP \qquad \textbf{[11.7]}$$
$$= AP(1+i)^2 + AP(1+i) + AP$$

In summary, the total future amount at the end of the nth period is

$$TA_{(n-1)} = AP(1+i)^{n-1} + AP(1+i)^{n-2} + \ldots + AP(1+i) + AP \qquad \textbf{[11.8]}$$

Note that this expression is a geometric series. One approach to finding the sum is described here.
Multiply both sides of Equation 11.8 by the factor $(1+i)$ to get:

$$(1+i)TA_{(n-1)} = AP(1+i)^n + AP(1+i)^{n-1} + \ldots + AP(1+i)^2 + AP(1+i) \qquad \textbf{[11.9]}$$

Subtracting both sides of Equation 11.8 from Equation 11.9 yields

$$[(1+i)TA_{(n-1)}] - TA_{(n-1)} = AP(1+i)^n - AP \qquad \textbf{[11.10]}$$

Rearranging Equation 11.10 results in

$$FA = TA_{(n-1)} = \frac{AP[(1+i)^n - 1]}{i} \qquad \textbf{[11.11]}$$

where FA is the future total amount.

Example 11.3 | **A** company purchases a mechanical system. Its expected maintenance cost at the end of each year is estimated to be $2,000. The expected useful life of the system is 15 years and the estimated annual interest rate is 7 percent. Calculate the future amount of the maintenance cost for the entire useful life of the system.

Solution

Substituting the specified data into Equation 11.11 yields

$$FA = (2,000)\frac{[(1+0.07)^{15}-1]}{(0.07)}$$

$$= \$50,258.04$$

Thus, the future value of the total maintenance cost is $50,258.04

11.4   PRESENT VALUE: UNIFORM PERIODIC PAYMENTS

The formula for the present value of uniform periodic payments is developed in a manner similar to that for the formula for the future value of uniform periodic payments.

$$PV_{pp} = AP[\frac{1-(1+i)^{-n}}{i}] \qquad \textbf{[11.12]}$$

where PV_{pp} is the present value of the uniform periodic payments. A detailed derivation of Equation 11.12 is given in Reference [2], and several other similar formulas may be found in Reference [3].

C alculate the present value of the total maintenance cost of the system described in Example 11.3. | **Example 11.4**

Solution

Using the Example 11.3 data in Equation 11.12, we get

$$PV_{pp} = (2000)\left[\frac{1-(1+0.07)^{-15}}{(0.07)}\right]$$

$$= \$18,215.83$$

The present value of the mechanical system total maintenance cost is $18,215.83.

11.5   ENGINEERING EQUIPMENT DEPRECIATION

The term *depreciation* means a decline in value. Engineering systems generally lose value with age. Four types of factors are considered responsible for this decrease in value: physical, functional, technological, and monetary. The physical factor represents operational wear and tear, which reduces the product's capability to perform its stated function. The functional factor represents a change in the demand for a product during its useful life, which makes it appear less valuable, even though it still functions effectively. The technological factor represents the development of better techniques, which make existing designs uneconomical or less efficient by comparison. The monetary factor represents finacial depreciation as the result of a change in the purchasing power of money.

Many design related decisions require an understanding of the approaches used to determine equipment depreciation. Several depreciation calculation methods have been developed. Three of these methods are described in the following sections [2, 4, 5].

11.5.1 STRAIGHT-LINE DEPRECIATION METHOD

Straight-line depreciation is probably the simplest and easiest approach. The method calls for setting aside an equal sum of money per year during the product's life span. The annual depreciation charge is expressed as

$$ADC = \frac{PC - SV}{m} \qquad \text{[11.13]}$$

where

ADC is the annual depreciation charge.

PC is the purchase cost of the item.

SV is the salvage value of the item.

m is the useful life of the item, in years.

The book value (or actual value) of the item at the end of year i is

$$BV_i = PC - i(ADC) \qquad \text{[11.14]}$$

where

i is the number of years in service.

BV_i is the product book value at the end of year i.

Example 11.5 | **A**ssume that an item of engineering equipment under design has an estimated selling price of $3 million, and its salvage value is $300,000. The predicted useful life of the equipment is 9 years. For a constant annual depreciation, determine the yearly depreciation charge for the product under design and its book value at the end of five years.

Solution

Substituting the given data into Equation 11.13 yields

$$ADC = \frac{[(3,000,000) - (300,000)]}{9}$$

$$= \$300,000$$

Similarly, inserting this result and the given data into Equation 11.14, we get

$$BV_5 = (3,000,000) - (5)(300,000)$$
$$= \$1.5 \text{ million}$$

Thus, the annual depreciation charge and the equipment book value at the end of five years are $300,000 and $1.5 million, respectively.

11.5.2 SUM-OF-THE-YEARS-DIGITS DEPRECIATION METHOD

The name of this method is derived from the calculation procedure utilized. The method allows a larger annual depreciation during the initial useful years of a product, and less in the later years. The annual depreciation is expressed as [6]

$$ADC = [PC - SV]\left[\frac{m - i + 1}{(1 + 2 + 3 + \cdots + m)}\right] \quad \textbf{[11.15]}$$

$$= (PC - SV)\left[\frac{2(m - i + 1)}{m(m + 1)}\right]$$

The book value of the product at the end of year i is

$$BV_i = 2(PC - SV)\left[\frac{\{1 + 2 + 3 + \cdots + (m - i)\}}{m(m + 1)}\right] + SV \quad \textbf{[11.16]}$$

For example 11.5 data, use the sum-of-the-years-digits depreciation method to calculate the equipment book value at the end of five years.

Example 11.6

Solution

Substituting the given data of Example 11.6 into Equation 11.16, we get

$$BV_5 = 2[(3,000,000) - (300,000)]\left[\frac{1 + 2 + 3 + \cdots + (9–5)}{9(9 + 1)}\right]$$

$$= \$600,000$$

The book value of the equipment at the end of the 5-year period will be $600,000.

11.5.3 DECLINING-BALANCE DEPRECIATION METHOD

The declining-balance method is also referred to as the constant percentage method or the Matheson formula. Two important characteristics of the method are: the write-off is accelerated in the initial years of the product service life, and the annual depreciation cost is a fixed percentage of the book value at the beginning of the year. Also, the method always assumes a positive salvage value for the equipment. The depreciation rate (factor) is expressed as

$$D_r = 1 - \left(\frac{SV}{PC}\right)^{1/m} \qquad \textbf{[11.17]}$$

At the end of the ith year, the book value BV_i of the equipment is

$$BV_i = PC[1 - D_r]^i \qquad \textbf{[11.18]}$$

Substituting Equation 11.17 into Equation 11.18 leads to

$$BV_i = PC\left[\frac{SV}{PC}\right]^{i/m} \qquad \textbf{[11.19]}$$

The annual depreciation charge ADC is given by

$$ADC = [BV_{i-1}]D_r \qquad \textbf{[11.20]}$$

Inserting for D_r from Equation 11.17, we get

$$ADC = [BV_{i-1}]\left[1 - \left(\frac{SV}{PC}\right)^{1/m}\right] \qquad \textbf{[11.21]}$$

Example 11.7

For example 11.5 data, use the declining-balance depreciation method to determine the equipment book value at the end of five years. Comment on the result in comparison to the one obtained in Example 11.6 using the sum-of-the-years-digits depreciation method.

Solution

Substituting the Example 11.5 data into Equation 11.19 yields

$$BV_5 = (3,000,000)\left[\frac{300,000}{3,000,000}\right]^{5/9}$$
$$= \$834,767.80$$

Thus, the equipment book value at the end of year 5 will be \$834,767.80. This book value is significantly higher than the one obtained through the use of the sum-of-the-years-digits depreciation method.

11.6 LIFE-CYCLE COSTING

The life-cycle cost of a product is the sum of all costs (i.e., procurement and ownership) during its entire life span. The history of the life-cycle costing concept goes back to the 1960s when the term "life cycle costing" was first introduced in a document [7] prepared for the United States Department of Defense. Since then, several publications on the subject have appeared; many are listed in Reference [2].

To perform the life-cycle cost analysis of an item, information is required on such areas as [2]:

1. Yearly operation and maintenance costs.
2. Acquisition cost.
3. Salvage value.
4. Interest and inflation rates.
5. Installation cost.
6. Delivery cost.
7. Tax benefit.

Operational costs include those costs associated with materials, labor, energy, supplies, insurance, and so on.

11.6.1 LIFE-CYCLE COST STUDIES

The use of the life-cycle cost concept is increasing in the industrial sector. Reasons for this upward trend include: increasing competition, increasing cost of ownership, rising inflation, expensive products, and increasing cost-effectiveness awareness among product users.

Information on the life-cycle cost of an item may have several uses: comparing competing projects, preparing long-range plans and budgets, comparing logistics concepts, controlling an ongoing project, deciding on the replacement of aging equipment, selecting among competing bidders, etc.

A life-cycle cost study may involve a number of steps, such as those given here [8].

1. Estimate the useful operational life of the product under consideration.
2. Estimate all associated costs, including operation and maintenance costs.
3. Estimate the final salvage value of the product.
4. Subtract the salvage value from the ownership cost.
5. Determine the present value of the resulting amount of the preceding step.
6. Add the amount of Step 5 to the acquisition cost of the product, to determine its life-cycle cost.
7. Repeat the above steps for all items being considered.
8. Compare the life-cycle costs of items, as desired.
9. Procure the product with the lowest life-cycle cost.

11.6.2 ESTIMATION MODELS

Over the years, various life-cycle cost estimation models have been developed. Some are general in nature, while others are tailored to a specific item. The objective of all models is to estimate the life-cycle cost of an item. In this section, we present two life-cycle cost models: one general, and the other specific.

General Model   This model is expressed by the following equation:

$$LCC = C_r + C_{nr} \qquad \text{[11.22]}$$

where

   LCC is the life-cycle cost of an item.

   C_r is the recurring cost associated with the item.

   C_{nr} is the nonrecurring cost associated with the item.

In turn, the recurring cost associated with the item is defined as

$$C_r = MC + OC + IC + SC + MPC \qquad \text{[11.23]}$$

where

   MC is the maintenance cost.

   OC is the operating cost.

   IC is the inventory cost.

   SC is the support cost.

   MPC is the manpower related cost.

Similarly, an item's nonrecurring cost is given as

$$C_{nr} = \sum_{i=1}^{10} C_i \qquad \text{[11.24]}$$

where C_i is the ith nonrecurring cost, defined as follows:

   $i = 1$,   procurement.

   $i = 2$,   support associated with the procurement action.

   $i = 3$,   training.

   $i = 4$,   research and development.

   $i = 5$,   installation.

   $i = 6$,   qualification approval.

   $i = 7$,   improving reliability and maintainability.

   $i = 8$,   transportation.

   $i = 9$,   life-cycle cost management.

   $i = 10$,   test equipment.

Specific Model   This specific model was developed to estimate the life-cycle cost of switching power supplies [9]. The switching power supply's life-cycle cost is expressed as

$$LCC_{sp} = C_i + C_f \qquad \text{[11.25]}$$

where

C_i is the initial cost.

C_f is the failure cost.

The cost of failure C_f is defined as

$$C_f = \lambda T (SC + RC) \qquad [11.26]$$

where

λ is the unit constant failure rate.

T is the expected life.

SC is the spare cost.

RC is the repair cost.

The spare cost in turn is given as

$$SC = \alpha(USC) \qquad [11.27]$$

where

α is the fraction of spares for each active unit.

USC is the unit spare cost.

11.6.3 LIFE-CYCLE COSTING AND EQUIPMENT SELECTION

The use of life-cycle costing in the selection of equipment is demonstrated by the following example.

A government department is considering purchasing a heavy-duty pump and three manufacturing companies A, B, and C, are bidding to provide the equipment. The department has various types of data available concerning the equipment, as shown in Table 11.1. The department management would like to know which company's pump is going to be the most suitable based on life-cycle cost.

Example 11.8

Solution: Company A's Pump

The annual expected cost FC_A of a failure is

$$FC_A = (\$15,000)(0.02)$$
$$= \$300$$

Thus, the present value PV_{MA} of the pump life-cycle maintenance cost is [2]

$$PV_{MA} = (\$300)\left[\frac{1 - (1 + 0.05)^{-15}}{0.05}\right]$$
$$= \$3113.90$$

Table 11.1 Pump data

No.	Description	Company A's Pump	Company B's Pump	Company C's Pump
1	Procurement cost	$400,000	$380,000	$420,000
2	Operating cost per year	$10,000	$12,000	$8,000
3	Useful life	15 years	15 years	15 years
4	Annual failure rate	0.02 failures per year	0.03 failures per year	0.01 failures per year
5	Cost of a failure	$15,000	$16,000	$9,000
6	Interest rate per year	5%	5%	5%

Similarly, the present value PV_{OA} of the pump life-cycle operating cost is

$$PV_{OA} = (\$10,000) \left[\frac{1 - (1 + 0.05)^{-15}}{0.05} \right]$$
$$= \$103,796.60$$

The life-cycle cost LCC_A of company A's pump is therefore

$$LCC_A = PC + PV_{MA} + PV_{OA}$$
$$= (\$400,000) + (\$3113.90) + (\$103,796.60)$$
$$= \$506,910.50$$

where PC is the pump procurement cost.

Solution: Company B's Pump

The annual expected cost FC_B of a failure is

$$FC_B = (\$16,000)(0.03) = \$480$$

Thus, the present value PV_{MB} of the pump life-cycle maintenance cost is

$$PV_{MB} = (\$480) \left[\frac{1 - (1 + 0.05)^{-15}}{0.05} \right]$$
$$= \$4,982.20$$

Similarly, the present value, PV_{OB}, of the pump life-cycle operating cost is

$$PV_{OA} = (\$12,000) \left[\frac{1 - (1 + 0.05)^{-15}}{0.05} \right]$$
$$= \$124,555.90$$

The life-cycle cost LCC_B of company B's pump is therefore

$$LCC_B = PC + PV_{MB} + PV_{OB}$$
$$= (\$380,000) + (\$4,982.20) + (\$124,555.90)$$
$$= \$509,538.10$$

Solution: Company C's Pump

The annual expected cost FC_C of a failure is

$$FC_C = (\$9{,}000)(0.01) = \$90$$

Thus, the present value PV_{MC} of the pump life-cycle maintenance cost is

$$PV_{MC} = (\$90)\left[\frac{1 - (1 + 0.05)^{-15}}{0.05}\right]$$
$$= \$934.20$$

Similarly, the present value PV_{OC} of the pump life-cycle operating cost is

$$PV_{OC} = (\$8,000)\left[\frac{1 - (1 + 0.05)^{-15}}{0.05}\right]$$
$$= \$83,037.30$$

The life-cycle cost LCC_C of company C's pump is therefore

$$LCC_C = PC + PV_{MC} + PV_{OC}$$
$$= (\$420,000) + (\$934.20) + (\$83,037.30)$$
$$= \$503,971.50$$

Solution: Summary

As a result of the analysis, we see that company C's pump would be the most economical to purchase, because its life-cycle cost is the lowest.

11.7 COST ESTIMATION MODELS

This section presents several cost estimation models taken from various published sources. These models would be useful in performing life-cycle cost studies, and comparing costs of items under study.

11.7.1 COST–CAPACITY MODEL

This model is used to approximate cost estimates where cost data are available for similar items of a different capacity. The cost of the desired item is estimated through the following relationship [2]:

$$C_d = C_{s0}\left(\frac{K_d}{K_{s0}}\right)^\theta \qquad \text{[11.28]}$$

where

C_d is the cost of the desired item.

K_d is the capacity of the desired item.

C_{s0} is the known cost of a similar item, of capacity K_{s0}.

θ is the cost capacity factor, the value of which varies for different equipment. Where no data are available, it is reasonable to assume the value of θ to be 0.6. Over the years, values of θ for various equipment items have been estimated [10–12]. Examples are: heat exchangers (0.6), pumps (0.6), heaters (0.8), conveyors (0.7), and tanks (0.7).

Example 11.9 | **A**ssume that the cost of a 50-passenger bus is $100,000. Estimate the cost of a similar 60-passenger bus if the value of the cost capacity factor is 0.6.

Solution

Substituting the given data into Equation 11.28 yields

$$C_d = (100,000)\left(\frac{60}{50}\right)^{0.6} = \$111,560.10$$

The cost estimate for the desired bus is $111,560.10

11.7.2   CORRECTIVE MAINTENANCE MODEL

This model is used to estimate corrective maintenance costs. The proposed equation is [2]

$$C_{cm} = C_m T_{sh}\left(\frac{MTTR}{MTBF}\right) \qquad\qquad \textbf{[11.29]}$$

where

$MTTR$ is the item's mean time to repair.

$MTBF$ is the item's mean time between failures.

C_m is the cost of maintenance labor per hour.

T_{sh} is the scheduled operating hours of the item.

Example 11.10 | **T**he mean time between failures and the mean time to repair of a mechanical pump are 500 hours and 20 hours, respectively. The pump's annual operating hours are 4,000 hours, and the hourly corrective maintenance labor cost is $25. Calculate the annual corrective maintenance labor cost of the mechanical pump.

Solution

Substituting the given data into Equation 11.29 yields

$$C_{cm} = (25)(4,000)\left(\frac{20}{500}\right) = \$4,000$$

The pump's annual corrective maintenance labor cost will be $4,000.

11.7.3 SOFTWARE MAINTENANCE MODEL

The software maintenance cost for maintaining the designed software is expressed as [12]

$$C_{sm} = 3(CPM)(ICH)/\alpha \qquad \text{[11.30]}$$

where

α is the difficulty constant.

CPM is the cost per man-month.

ICH is the number of instructions to be changed per month.

11.7.4 MACHINING MODEL

The machining cost for producing a new part or modifying an existing part to fit the new design is expressed as [11, 13, 14]

$$MC = rT = \frac{1}{60}\left[\frac{C_m(1+OR_m)+L_{0r}(1+OR_o)}{100}\right](MT+NT) \qquad \text{[11.31]}$$

where

T is the production time for a unit.

r is the cost rate per minute (dollars).

C_m is the machine cost per hour (dollars).

OR_m is the machine overhead rate (percentage).

L_{0r} is the operator labor rate per hour (dollars).

OR_0 is the operator overhead rate (percentage).

MT is the machining time.

NT is the nonproduction or idle time.

11.7.5 CUTTING TOOL MODEL

This model is useful for estimating the tool cost associated with a cuting tool brazed to the tool holder. The cutting tool cost C_{ct} is expressed as [11]

$$C_{ct} = [(RC)n + C_T]/(n + 1) \qquad \textbf{[11.32]}$$

where

n is the number of resharpenings.

C_T is the cost of the tool (dollars).

RC is the tool resharpening cost (dollars).

11.7.6   AC MOTOR OPERATION MODEL

Certain costs are associated with the operation of an AC motor. The following can be used to estimate this cost [15]:

$$MOC = \frac{(0.746)(HP_m)T_m C_e}{\alpha} \qquad \textbf{[11.33]}$$

where

MOC is the cost to operate the motor.

α is the motor efficiency.

C_e is the electricity cost per kilowatt-hour ($/KWH).

T_m is the motor operational time per year, expressed in hours.

HP_m is the motor horsepower.

Example 11.11

An air compressor has a 20-horsepower motor, and the yearly estimated operational time of the motor is 5,000 hours. Furthermore, the motor efficiency is 85 percent. Calculate the annual cost to operate the motor, if the cost of electricity is 2 cents per kilowatt-hour.

Solution

Substituting the given data into Equation 11.33 yields

$$MOC = \frac{(0.746(20)(5,000)(0.02)}{0.85}$$
$$= \$1,755.30$$

Thus, the annual cost to operate the motor will be $1,755.30.

11.7.7   LIGHTING MODEL

This model is used to estimate annual lighting costs and is expressed as [15]

$$LC = \frac{W_b T C_e}{1,000} \qquad \text{[11.34]}$$

where

    LC is the annual cost of lighting.

    C_e is the electricity cost per kilowatt-hour.

    W_b is the wattage of the lightbulb (watts).

    T is the bulb's annual operational time, expressed in hours.

A 100-watt lightbulb is operated for 4,000 hours in one year. Calculate the annual cost of the lighting, if the electricity cost is 2 cents per kilowatt-hour.   **Example 11.12**

Solution

Substituting the given data into Equation 11.34 leads to

$$LC = \frac{(100)(4,000)(0.02)}{1,000} = \$8$$

The cost of the lighting will be $8.

11.8 PROBLEMS

1. What is the difference between simple and compound interest?
2. Develop an expression for the present value of uniform periodic payments.
3. Compare the following depreciation methods:

   Straight-Line.

   Sum-of-the-Years-Digits.

   Declining-Balance.
4. The estimated selling price and the salvage value of an engineering system under design are $10 million and $0.5 million, respectively. The expected useful life of the system is 12 years. For constant yearly depreciation, calculate the annual depreciation charge for the system and its book value at the end of eight years.
5. Define the term *life-cycle costing* and give reasons for its inception.
6. Outline the steps for conducting a life-cycle costing study.
7. An engineering organization is considering procuring a mechanical system for use in its production line. Two manufacturing companies A and B are competing

to provide the system. The data available on both the systems are given in Table 11.2. Determine which of the two systems should be purchased, using life-cycle cost.

Table 11.2 Data for mechanical systems

No.	Description	Company A's System	Company B's System
1	Acquisition cost	$2 million	$2.4 million
2	Annual interest rate	7%	7%
3	Annual failure rate	0.04 failures/year	0.03 failures/year
4	Cost of a failure	$25,000	$20,000
5	Useful life	12 years	12 years
6	Annual operating cost	$31,000	$30,000

8. Assume that an automobile is used for 3,000 hours per year and its mean time between failures is 400 hours. The automobile's mean time to repair is 6 hours and the hourly corrective maintenance labor cost is $30. Compute the yearly labor cost for corrective maintenance on the automobile.

9. An electric 40-horsepower motor is operated for 4,000 hours annually. The motor efficiency is 80 percent and the cost of electricity is 3 cents per kilowatt-hour. Calculate the annual cost to operate the motor.

REFERENCES

1. DeGarmo, P.E.; J.R Canada; and W.G. Sullivan. *Engineering Economy.* New York: Macmillan, 1979.

2. Dhillon, B.S. *Life Cycle Costing.* New York: Gordon and Breach Science Publishers, 1989.

3. Fabrycky, W.J; and B.S. Blanchard. *Life Cycle Cost and Economic Analysis.* Englewood Cliffs, NJ: Prentice-Hall, 1991.

4. Dieter, G.E. *Engineering Design.* New York: McGraw-Hill 1983.

5. Walton, J.W. *Engineering Design.* New York: West Publishing Company, 1991.

6. Riggs, J.L. *Economic Decision Models for Engineers and Managers.* New York: McGraw-Hill, 1968.

7. *Life Cycle Costing in Equipment Procurement.* Report No. LMI Task 4C-5. Washington, D.C.: Logistics Management Institute (LMI), April 1965.

8. Coe, C.K. "Life Cycle Costing by State Governments." *Public Administration Review,* September/October 1981, pp. 564–569.

9. Nelson, H.C.; O.P. Kharbanda; W.A. Janda; and J.H. Black. *Capital Investment Cost Estimation in Cost and Optimization Engineering,* ed. F.C. Jelen, and J.H. Black. New York:McGraw-Hill, 1983, pp. 321–381.

10. Desai, M.B. "Preliminary Cost Estimating of Process Plants." *Chemical Engineering,* July 1981, pp. 65–70.

11. Dieter, G.E. *Engineering Design.* New York: McGraw-Hill, 1983, pp. 324–366.

12. Sheldon, M.R. *Life Cycle Costing: A Better Method of Government Procurement.* Boulder, CO: Westview Press, 1979.

13. Armarego, E.J.A.; and R.H Brown. *The Machining of Metals.* Englewood Cliffs, NJ: Prentice-Hall, 1969.

14. Boothroyd, G. *Fundamentals of Metal Machining and Machine Tools.* New York: McGraw-Hill, 1975.

15. Brown, R.J; and R.R Yanuck. *Life Cycle Costing: A Practical Guide for Energy Managers.* Atlanta, GA: Fairmont Press, 1980.

chapter

12

MATERIALS SELECTION, MANUFACTURING, AND ENVIRONMENTAL DESIGN

12.1 INTRODUCTION

In our day-to-day activities, we come into contact with a wide variety of materials in many distinct forms, shapes, sizes, or applications. To varying degrees, these materials affect human comfort, progress, safety, etc. The design of an engineering product must carefully consider the selection of the materials to be used. This task may not be as easy as it might appear, as there are over 40,000 metallic alloys and a very large number of nonmetallic materials for engineering products [1]. An incorrect selection may result in poor product reliability or an unnecessarily costly product, because in some manufacturing operations the materials cost may be over 50 percent of the total cost. In the automobile and shipbuilding industries, for example, the cost of materials accounts for approximately 70 percent and 45 percent, respectively, of the manufacturing cost.

Manufacturing is the economic term for making goods accessible to fulfill human needs. The ultimate objective of any engineering product design is to translate ideas into physical form as economically as possible, without sacrificing its effectiveness. This means that the manufacturing aspects of the design under consideration must be carefully considered, by the design engineer and/or by the manufacturing engineer working with the design engineer. Any product design that cannot be physically produced effectively is as bad as no design at all. Typically, the manufacturing cost is the largest component of the selling price of a product. It accounts for approximately 40 percent of a product's selling price [2]; in turn, parts and materials are responsible for roughly 50 percent of the manufacturing cost.

One problem with the advances in technology is the deterioration in environment. In some places of the world, the environmental problems, such as smog and ozone layer depletion, are very serious. Many governments have passed various legislation to protect citizens, wildlife, etc. Therefore, new engineering designs are expected to consider carefully the effects on the environment.

This chapter addresses material selection, manufacturing, and the environment as they affect engineering designs.

12.2 MATERIALS SELECTION

During the design process, one of the crucial tasks is the selection of the materials and parts to be used. This requires careful consideration by the design personnel, as materials and parts come in numerous shapes, sizes, classifications, compositions, etc. An error in the selection process may result in an unsuccessful end product.

12.2.1 MATERIAL CLASSIFICATIONS AND PROPERTIES

The various types of materials that may be used in a product design are classified as follows [3]:

1. **Metals.** This important materials classification can be further divided into two categories: ferrous and nonferrous alloys. Ferrous alloys are based on iron, which has good mechanical properties. Nonferrous alloys are based on materials other than iron, such as copper, tin, aluminum, and lead.

2. **Ceramics and glass.** These are the result of the combination of metallic and nonmetallic elements. They are: good insulators (heat and electricity), brittle, thermally stable, more wear resistant compared to metals, harder, and lower in thermal expansion than most metals.

3. **Woods and organics.** These are obtained from trees and plants. Their major advantage is that they are a renewable resource. However, some of their drawbacks are that they absorb water, require special treatment to prevent rotting, and are more flammable.

4. **Polymers or plastics.** These materials change viscosity with variations in temperature; therefore, they are easy to mold into a given shape. In comparison to metals, plastics have a deficiency of free electrons in their atomic structures. There are several benefits of polymers: good insulators (heat and electricity); resistant to chemicals and water; smooth surface finish; available in many colors, eliminating the need for painting. There are also several drawbacks of polymers: low strength, dimensional instability, deterioration in ultraviolet light, and excessive creep at all temperatures.

The properties of materials can be divided into six categories [1, 4]: mechanical, thermal, physical, chemical, electrical, and fabrication. Mechanical properties include: fatigue, strength, elasticity, wear, hardness, and plasticity. Thermal properties include: absorptivity, conductivity, fire resistance, and expansion coefficient. Physical properties are: density, permeability, viscosity, crystal structure, porosity, and dimensional stability. Chemical properties include: corrosion, oxidation, hydraulic permeability, and biological stability. Four important electrical properties are: hysteresis, conductivity, coercive force, and dielectric constant. There are several fabrication properties, some of which are: weldability, castability, machinability, and heat treatability.

12.2.2 MATERIALS SELECTION PROCESS

Over the years, several systematic approaches have been developed for selecting materials. One of those procedures, consisting of four steps, is given here.

1. **Perform a material requirements analysis.** This step is concerned with determining the service and environment conditions the product will have to endure.

2. **Select promising materials.** This step calls for filtering through available materials to obtain several possible suitable candidates for the intended application.

3. **Choose the most suitable material.** This step involves analyzing the promising materials with respect to such factors as cost, performance, availability, and fabrication ability and then selecting the most suitable material.

4. **Obtain the necessary test data.** This final step is concerned with experimentally determining the important properties of the selected material, to obtain statistically reliable performance measures under real-life operational conditions.

It should be emphasized that there are many factors to consider during the materials selection process. In addition to those already mentioned, other selection factors include the following [4]:

1. **Product specification fulfillment.** The selected material must satisfy the stated specifications.

2. **Cost**. This important factor plays a dominant role in marketing of the end product. The cost effectiveness of the recommended materials must be carefully evaluated by the design professional.

3. **Material availability**. This involves determining the availability of the recommended materials at the desired time frame at an acceptable cost. The evaluation of material availability must consider legal, political, and competitive factors.

4. **Material joining approaches**. Under real-life conditions, it may be impractical to produce an item using a single piece of material. This may call for manufacturing the components with different pieces of material joined together to form a single item or unit. Three important materials joining approaches are: metallurgical processes, adhesive joining, and mechanical approaches. Soldering, welding, and brazing are examples of metallurgical joining processes. Adhesive bonding is a relatively recent development that uses polymer resins. However, this approach is generally considered less reliable and less durable than metallurgical joinings. There are several mechanical methods used for joining, including: rivets, nuts and bolts, self-tapping screws, etc.

5. **Fabrication**. Engineering products usually require some level of fabrication, and many different fabrication techniques are available. Factors affecting the fabrication method selected include: time constraints, material type, product application, cost, quantity to be manufactured, etc. The design professional must decide whether to select the material first and then specify the suitable fabrication methods, or vice-versa.

6. **Technical issues**. Technical factors mainly concern a material's mechanical properties. Examples include: strength compared to anticipated load, safety margin, weight, temperature variation, and potential loading changes.

7. **Testing**. This factor addresses the type of testing to be required, the frequency of testing, the amount of testing, etc.

12.2.3 SELECTED PUBLISHED LITERATURE

There are many excellent publications on materials and related topics, all of which should provide useful information to design professionals. Some of these are as follows:

1. Chew, R.Z. *Materials Engineering*. Cleveland, OH: Penton, 1988.
2. Higdon, A. *Mechanics of Materials*. New York: Wiley, 1985.
3. *Metals Handbook, Desk Edition*. ed. H.E. Boyer, and T.L. Gall. Metals Park, OH: American Society for Metals, 1984.
4. *Encyclopedia of Materials Science and Engineering*. ed. M.B. Bever. Cambridge: The MIT Press, 1986.
5. Budinski, K. *Engineering Materials Properties and Selection*. Englewood Cliffs, NJ: Prentice-Hall, 1989.
6. Lankford, W.T., et al. *The Making, Shaping, and Treating of Steel*. Pittsburgh: Association of Iron and Steel Engineers, 1985.
7. Landrock, A.A. *Adhesives Technology Handbook*. Park Ridges, NJ: Noyes Publications, 1985.
8. *Welding Handbook* 3. New York: American Welding Society, 1976.
9. Davis, H.E.; G.E. Troxell; and G.F.W. Hauck. *The Testing of Engineering Materials*. New York: McGraw-Hill, 1982.
10. Collins, J.A. *Failure of Materials in Mechanical Design*. New York: Wiley, 1981.
11. *Aluminum: Properties and Physical Metallurgy*. ed. J.E. Hatch. Metals Park, OH: American Society for Metals, 1984.
12. *Titanium Alloys Handbook*. Columbus, OH: Metals and Ceramics Information Center, Battelle Columbus Laboratories, 1972.
13. Barry, B.T.K.; and C.G. Thwaites. *Tin and its Alloys and Compounds*. New York: Wiley, 1983.
14. Juran, R. *Modern Plastics Encyclopedia*. New York: McGraw-Hill, 1989.
15. *Source Book on Materials for Elevated Temperature Applications*. Metals Park, OH: American Society for Metals, 1979.
16. *Steel Castings: Handbook*. Cleveland, OH: Steel Founder's Society of America, 1980.

17. Schwartz, M.M. *Metals Joining Manual.* New York: McGraw-Hill, 1979.

18. *Handbook of Glass Manufacture.* ed. F.V. Tooley. New York: Ashlee, 1984.

19. *Wood Engineering Handbook.* Englewood Cliffs, NJ: Prentice-Hall, 1982.

20. *Standard Handbook of Fastening and Joining.* ed. R.E. Parmley. New York: McGraw-Hill, 1977.

21. Kirby, G.N. "How to Select Materials." *Chemical Engineering,* November 1980, pp. 86–149.

22. *Source Book on Failure Analysis.* Metals Park, OH: American Society for Metals, 1975.

23. *Ceramic Source.* Columbus, OH: American Ceramic Society, 1989.

24. Redmond, J.D.; and K.H. Miska. "The Basics of Stainless Steels." *Chemical Engineering,* October 1982, pp. 78–118.

25. Naumann, A. *Failure Analysis: Case Histories and Methodology.* Metals Park, OH: American Society for Metals, 1983.

26. Graff, G.M. "Engineering Plastics." *Chemical Engineering,* August 1982, pp. 42–45.

27. Hanley, D.P. *Introduction to the Selection of Engineering Materials.* New York: Van Nostrand Reinhold, 1980.

28. Harvey, P.D. *Engineering Properties of Steels.* Metals Park, OH: American Society for Metals, 1982.

29. Clauser, H.R.; R.J. Fabian; and J.A. Mack. "How Materials are Selected." *Materials in Design Engineering,* July 1965, pp.109–128.

30. Campbell, J.E., et al. *Application of Fracture Mechanics for Selection of Metallic Structural Materials.* Metals Park, OH: American Society for Metals, 1982.

31. Farag, M.M. *Materials and Process Selection in Engineering.* London: Applied Science Publishers, 1979.

12.2.4 ENGINEERING PRODUCT MATERIALS

Various engineering products commonly used in day-to-day life use different types of materials. The materials used in selected common engineering products are presented here [5]:

1. **Bearings and brushings,** which use such materials as bronze, teflon®, stainless steel 316, nylon, and beryllium.

2. **Valves,** which use such materials as nylon, aluminum, bronze, acetal and nylon.

3. **Heat exchangers,** which use stainless steel.

4. **Auto engines,** which are made of grey cast iron.

5. **Electric connectors**, which use such materials as nylon, phenolic, polyethersulfone, and polysulfone.

6. **Springs**, which use such materials as stainless steel, maraging steel, steel 1080, acetal, and beryllium copper.

7. **Fan blades**, which are made of steel 4340 and nylon.

8. **Conveyor chains**, which are made of steel 1040 and acetal.

9. **Bolts and screws**, which use such materials as nylon, steel 1020, steel 1040, steel 4140, and acetal.

10. **Gears**, which use such materials as grey iron, aluminum bronze, polyamide, acetal, steel 1020, and nylon.

12.2.5 CASE STUDY: PETROLEUM HEAT EXCHANGER FAILURE

A petroleum heat exchanger was used in Kenai, Alaska, to reheat offshore petroleum to lower its viscosity so that it could be effectively pumped further inland to storage facilities [6]. An accident lead to a raging fire in that heat exchanger. A subsequent investigation revealed that the cause of the accident was the development of a crack in one of the U-tube joints, which allowed the ingestion of petroleum into the flame tube and the point of combustion. The accident was specifically the result of a design deficiency, which was corrected in a subsequent redesign.

12.3 MANUFACTURING

Any good design can only be successful if the product is properly manufactured. Manufacturing aspects must therefore be carefully considered during the design phase. Manufacturing engineers are often part of the design team, depending on the nature of the product, the philosophy of the management, and the complexity of the manufacturing process. This section discusses the basic concepts of manufacturing.

12.3.1 MANUFACTURING SYSTEM CATEGORIES

Manufacturing systems may be grouped into several major categories [2]: job shop, project shop, flow-line, continuous processes, and cellular shop. The first four classifications are the classic systems, whereas the last category, cellular shop, is a new kind of system. The job shop involves the manufacture of various types of products leading to small production lot sizes. This manufacturing system requires a relatively high worker skill level for carrying out a variety of tasks, as well as general purpose production equipment. Examples of job shop products include: machine tools, space vehicles, and special tools. Hospitals, auto repair shops, universities, metal fabrication shops, and machine shops are examples of the job shop manufacturing system.

In the project shop manufacturing system, the product remains in a fixed location during the manufacturing process, due to such factors as product size and weight. The resources, including people, machines, and materials, are brought to the product location. Examples of such products are: locomotives, large aircraft, ships, houses, bridges, and television shows.

The flow-line manufacturing system is used for very high production rates and a product-inclined layout. This requires the installation of specialized manufacturing equipment geared to a specific product. Examples of the flow-line manufacturing system are: cafeteria, automobile assembly line, car wash, and television set assembly line. Most factories combine the job shop and flow-line systems.

The continuous processes manufacturing system is used in situations where the product (for example, liquids, powders, and gases) physically flows. This system is the most efficient and the least flexible. Typical examples are: an oil refinery, an electric power generation plant, and a chemical plant.

The cellular shop is the newest manufacturing system and is made up of linked cells. The individual processes required to manufacture a product are grouped into cells according to sequence and operational requirements. The cellular shop system is similar to the flow shop or flow-line arrangement, but with added flexibility. Examples of the cellular arrangement are fast-food restaurants, robotic cells, and Midas Muffler.

12.3.2 MANUFACTURING PROCESSES CATEGORIES AND SELECTION FACTORS

Many manufacturing processes have certain similarities and distinctions. A large number of such processes can be classified into eight major groups, as shown in Table 12.1 [1].

Table 12.1 Major classifications of manufacturing processes

No.	Classification
1	Polymer related processes
2	Machining processes
3	Assembly related processes
4	Material joining processes
5	Solidification associated processes
6	Thermal heat treatment and surface treatment related processes
7	Particulate processing
8	Deformation processes

Polymer related processes are strictly concerned with polymers and were developed because of the special properties of such materials. Two typical examples of these processes are thermoforming and injection melding. Machining processes are basically concerned with removing raw material with the use of a sharp tool. Some of the methods used for material removal are: grinding, milling, shaving, and turning.

Assembly related processes, as the name suggests, are concerned with bringing together or combining items to produce a finished product or a subassembly. Joining processes are concerned with joining materials or parts through brazing, riveting, bolting, welding, soldering, adhesive bonding, and so on. In solidification associated processes, molten metal, plastic, or glass is cast into a mold and solidified. Thermal heat treatment and surface treatment related processes are concerned with improving mechanical and surface properties. Particulate processing is concerned with consolidating metal, polymer, or ceramic particles through pressing and sintering, plastic deformation, or hot compaction, as well as composite material processing. Finally, in deformation processes, the materials (usually metals) are deformed while hot or cold, for the purpose of strengthening their properties and changing their shapes.

In the selection of a manufacturing process, the influencing factors include the following [1]:

1. Materials to be used.
2. Manufacturing cost.
3. Equipment availability.
4. Quantity required.
5. Tools and jigs available and required.
6. Delivery date.
7. Surface finish requirements.
8. Tolerance level requirements.

12.3.3 SELECTED MANUFACTURING OPERATIONS

The manufacture of an item requires various operations, of which the common ones are: casting, forging, machining, welding, and assembly. Each of these operations is described in the following paragraphs [1–3].

Casting Casting is a widely used first step in the manufacturing process. During casting, an item takes on its initial usable shape and chemistry. In the casting process a solid is melted down, heated to a desirable temperature level, and, if called for, treated to change its chemical composition. The molten material is then poured into a mold made in the required shape. The cast items may range in size and weight from a fraction of an inch to several yards and from an ounce to several tons. Typical examples are a zipper's individual teeth, propellers, and the stern frames of ships.

Table 12.2 lists widely known casting processes. Each of these processes is described in References 1 and 3. Factors such as weight, cost, shape complexity, required tolerance level, strength, quantity, required quality, and required finish surface dictate the selection of a casting process.

Finishing costs and scrap losses associated with die casting, shell casting, permanent-mold casting, and investment casting are low. In contrast, sand casting finishing costs are high and scrap loss is moderate. Some of the benefits of casting are:

1. Material is placed only where it is required.

2. Material removal is kept to a minimum for irregularly-shaped items.

3. Some items, such as boat anchors, wheel weights, and cabinet handles, require no machining whatsoever.

Table 12.2 **C**ommon casting methods

No.	Casting method
1	Die casting
2	Centrifugal casting
3	Shell mold casting
4	Investment casting
5	Ceramic mold casting
6	Sand mold casting
7	Permanent mold casting
8	Plaster mold casting
9	Full-mold casting

Some of the drawbacks of the casting are:

1. Materials must be melted down.

2. Solidification requires a certain time period.

3. Molds must be prepared.

4. Process cycle for making a pattern is relatively slow.

Forging   Forging, which is among the most important methods of manufacturing items for high-performance uses, involves changing the shape of a piece (blank) of material by exerting force on that blank. The methods of applying pressure include: mechanical press, hydraulic press, and drop hammer. Products such as crankshafts, wrenches, and connecting rods are the result of forging. The material to be forged may be hot or cold, i.e., above or below the crystallization temperature. Several characteristics associated with hot forging include [3]:

1. Requires less force to form the desirable shape.
2. Requires special heat resistant tools.
3. Deteriorates tools more rapidly than cold forging.
4. Does not change the hardness or ductility of the material.
5. Requires a cooling-off period prior to handling or machining operations.
6. Makes the material tougher, because the material grains become smaller.
7. Increases the process cycle time, because of the need to heat the material to the necessary forging temperature.
8. Increases the effectiveness of critical forming without rupturing the material.

Some of the characteristics of cold forging are:

1. Increases the material yield strength.
2. Increases the process efficiency compared to hot forging (because the material does not require heating).
3. Limits the amount of forging permitted, because of the danger of material rupture with excess deformation.
4. Increases the force needed to produce the desired shape.

Three well-known forging processes are: closed-die forging, upset forging, and cold heading. The tooling costs for each process are high, high, and medium, respectively, and the direct labor costs are medium, low, and low, respectively.

Machining   Machining involves removing unwanted material from a block (or blank) of material, according to given specifications, such as size, shape, and finish, to produce a finished product. Machining is the most important of the primary manufacturing processes. In the United States alone, $60 billion is spent per annum for metal removal operations [2].

There are many machining processes with which the design professional should be familiar. The most commonly used are: turning, milling, boring, grinding, drilling, and broaching. The typical production rates associated with each of these processes are: 1–10 parts/hour (engine lathes), 1–100 parts/hour, 2–20 hours/piece, 1–1000 pieces/hour, 2–20 seconds/hole after setup, and 300–400 parts/minute, respectively. The raw material forms associated with each of these rates are: cylinders, forgings, preforms, and castings; rods, bars, tubes, and plates; preforms and castings; rods, bars, and plates; bars, preforms, and plates; and rods, plates, tubes, and bars, respectively. The maximum and minimum sizes applicable to each machining process category are given in Reference 2.

Welding   Welding is a versatile production process which is used for combining items produced by some other means of manufacturing. Welding is the process of permanently joining two materials through coalescence, which involves a combination of pressure, temperature, and surface conditions. The American Welding Society has categorized common welding processes into five major areas, as follows [2]:

1. Resistance welding.

2. Arc welding.

3. Oxyfuel welding.

4. Unique welding.

5. Solid-state welding.

The three types of resistance welding are: resistance spot welding, projection welding, and resistant seam welding. Arc welding encompasses seven processes: flux cored arc welding, gas tungsten arc welding, shielded metal arc welding, stud welding, plasma arc welding, submerged arc welding, and gas metal arc welding. The oxyfuel gas welding classification includes pressure gas welding and oxyacetylene welding. The six types of unique welding are: induction welding, laser beam welding, thermit welding, electron beam welding, flash welding, and electroslag welding. Finally, solid-state welding covers six classifications: roll welding, forge welding, ultrasonic welding, cold welding, explosion welding, and friction welding.

Some of the important characteristics of welding are:

1. X-ray inspection equipment can be used to verify weld integrity.

2. Dissimilar materials can be joined.

3. Welding equipment is portable.

4. Special skills are required to perform overhead welding.

5. Metal heat treatment can occur.

6. Heat distortion can occur.

7. Shapes can be created.

8. The possibility of loose part vibration is eliminated.

Some of the design related guidelines associated with welding are as follows [7]:

1. Weld equal-thickness parts together.

2. Reduce the number of welds in an item.

3. Provide adequate ventilation to remove the resulting gases and smoke.

4. Locate the welds where loads or deflections, or both, are not crucial.

5. Where possible, avoid the use of welded laps, stiffeners, and straps.

6. Where possible, avoid overhead welding; design for welding in the horizontal position.

7. Provide sufficient room for welders and their equipment to work.

8. Develop a sequence to be followed in welding various parts together.

9. Provide adequate training and equipment to the welders.

Reference 2 categorizes the common welding processes into three groups according to rate of heat input, as follows:

1. **High**. This includes such welding processes as laser, percussion, spot and seam resistance, plasma arc, and electron beam.

2. **Moderate**. This includes five welding processes: gas tungsten arc, shielded metal arc, gas metal arc, submerged arc, and flux cored arc.

3. **Low**. Three welding processes belong in this category: flash, oxyfuel, and electro-slag.

The high-heat-input welding processes tend to generate low total heat content within the metal, small heat-affected zones, and fast cooling rates. On the other hand, the low-heat-input processes generate high total heat, large heat-affected areas, and slow cooling rates.

Assembly   Assembly processes combine individual items into an end product such that the product can be disassembled, if needed, without damaging the parts. Two important factors to be considered for this process are cost and quality. In general, assembly processes can be grouped into the two major areas shown in Table 12.3 [1].

Table 12.3 Classification of component assembly processes

No.	Classification Name	Description
1	Automatic	
	Type I	Involves robots and parts magazines.
	Type II	Uses automatic feeders and special-purpose synchronous indexing machines.
	Type III	Uses automatic feeders, special-purpose free transfer, and nonsynchronous machines.
	Type IV	Uses parts magazines, special-purpose free transfer, and nonsynchronous machines with programmable work heads.
2	Manual	
	Type I	Basically involves manual operations.
	Type II	Uses parts feeders and is known as mechanically aided manual assembly.

The following factors play an important role in deciding between manual and automatic assembly:

1. Number of assembled units to be produced.
2. Design complexity.
3. Capital and labor relative costs.

4. Parts per assembly.

5. Component complexity.

For more information on assembly processes, consult References 8 and 9.

12.3.4 SELECTED MANUFACTURING LITERATURE

Some of the many excellent publications and published literature on manufacturing are listed here.

1. DeFazio, T.L.; and D.E. Whitney. "Simplified Generation of Mechanical Assembly Sequences." *IEEE Journal of Robotics and Automation* 3, no.6, (1987), pp. 110–115.

2. Amrine, H.T.; J.A. Ritchey; C.L. Moodie; and J.F. Kmec. *Manufacturing Organization and Management.* Englewood Cliffs, NJ: Prentice-Hall, 1993.

3. Black, J.T. "Cellular Manufacturing Systems—An Overview." *IIE Journal,* November 1983, p. 36.

4. Schonberger, R.T. *Japanese Manufacturing Techniques.* New York: Free Press, 1982.

5. *Tool and Manufacturing Handbook.* ed. D.B. Dallas. New York: McGraw-Hill, 1976.

6. Amstead, B.H.; P.F. Ostwald; and M.L. Begelman. *Manufacturing Processes.* New York: Wiley, 1979.

7. Dorf, R.C. *Robotics and Automated Manufacturing.* Englewood Cliffs, NJ: Prentice-Hall, 1984.

8. Whitney, D.E. "Manufacturing by Design." *Harvard Business Review,* July-August 1988, pp. 83–91.

9. *Machining Data Handbook.* Cincinnati, OH: Metcut Research Associates, 1980.

10. Hitomi, K. *Manufacturing Systems Engineering.* London: Taylor & Francis Ltd., 1979.

11. Ostwald, P.R. "Manufacturing Cost Estimating Guide." *American Machinist,* 1982, pp. 80–81.

12. Weintraub, C. "Design for Manufacturability." *Design Graphics World,* October 1983, pp. 14–16.

13. Dhillon, B.S. *Life Cycle Costing.* New York: Gordon and Breach Science Publishers, 1989, Chapter 6.

14. DeGarmo, E.P.; J.T. Black; and R.A. Kohser. *Materials and Processes in Manufacturing.* New York: Macmillan, 1988.

15. Schey, J.A. *Introduction to Manufacturing Processes.* New York: McGraw-Hill, 1977.

16. *Welding Handbook.* New York: American Welding Society, 1968.

17. Heine, R.W.; C.R. Loper; and C. Rosenthal. *Principles of Metals Casting.* New York: McGraw-Hill, 1967.

12.3.5 SELECTED MANUFACTURING MODELS

This section presents a selected number of mathematical models developed for manufacturing processes [2]. The models discussed are: metal removal rate model, estimation model, drilling cost estimation model, tapping cutting time estimation model, and the turning operation cost estimation model.

Block Metal Removal Rate (MRR) Model The block metal removal rate (MRR) model is used to estimate the volume of metal removed from a blank per unit time. The equation is

$$MRR = \frac{VC}{T}\,(cubic\ in.\ per\ minute) \qquad [12.1]$$

$$VC = LMN \qquad [12.2]$$

where

VC is the volume removed.

T is the time taken for VC expressed in minutes.

L is the length of block being cut.

M is the depth of the cut.

N is the width of the block being cut.

Metal Removal Rate (MRR) for Turning Model This model is concerned with estimating the metal removal rate for turning operations. In this case, the MRR is

$$MRR = 12(VL_c)F(d_u^2 - d_c^2)/4d_u \qquad [12.3]$$

where

VL_c is the cutting velocity.

F is the feed rate, expressed in inches per revolution.

d_u is the uncut diameter of the piece.

d_c is the cut diameter of the piece.

Drilling Time Estimation Model This model is used to estimate drilling time required per hole, T_h, and is expressed as

$$T_h = \frac{L + A}{F_r} + \frac{L_T}{R} + T_p \qquad [12.4]$$

where

T_p is the downtime required to change the drill prorated per hole.

L_T is the rapid-traverse length, including withdrawal.

R is the rate of rapid traverse.

L is the drilled distance.

A is the air cut.

F_r is the feed rate.

The feed rate F_r is given by

$$F_r = F(RV) \qquad [12.5]$$

where

F is the feed, expressed in inches per revolution.

RV is the revolutions per minute.

The downtime required to change the drill, prorated per hole is given by

$$T_p = \frac{T_d}{H} \qquad [12.6]$$

where

T_d is the drill change downtime.

H is the number of holes drilled per drill regrind.

Drilling Cost Estimation Model   This model is used to estimate the cost per hole drilled C_h and is expressed as

$$C_h = DT(LC + MR) + PK \qquad [12.7]$$

where

DT is the drilling time per hole.

LC is the labor cost.

MR is the machine rate.

PK is the cost of buying and regrinding the drill, prorated per hole.

The cost PK of buying and regrinding the drill, prorated per hole, is

$$PK = (BC + GC)N_{rg}/N_h. \qquad [12.8]$$

where

BC is the buying cost.

GC is the cost per grind.

N_{rg} is the number of regrinds.

N_h is the number of holes drilled.

Tapping Cutting Time Estimation Model This model approximates the tapping cutting time and is expressed as

$$TC \approx \pi(DT)D_t\beta/8S \qquad \text{[12.9]}$$

where

S is the cutting speed.

β is the number of threads per inch.

D_t is the tapped hole depth, expressed in inches.

DT is the diameter of the tap, expressed in inches.

Turning Operation Cost Estimation Model This model is used to determine the total cost per piece for a turning operation and is expressed as

$$TK = MC + HC + TCH + TC \qquad \text{[12.10]}$$

where

TK is the total cost per piece for the turning operation.

TC is the cost of the tool.

HC is the handling cost.

MC is the cost of machining.

TCH is the cost of tool changing.

12.4 ENVIRONMENTAL DESIGN

Today, an engineering designer must carefully evaluate the potential impact of a design on the environment. In many cases, governmental or other environmental guidelines must be fully satisfied by the product. A number of well-publicized engineering system failures, for example, the Chernobyl nuclear accident, and the Alaska oil spill, have created a new awareness of environmental concerns to be addressed during the design phase.

In the late 1960s, the United States Congress began considering legislation to protect the environment. As a result of this effort, the National Environmental Policy Act (NEPA) and the Clean Air Act were passed, and the Environmental Protection Agency (EPA) was created. Since its creation, this agency has made an excellent effort to protect the nation's environment. However, many problems [10] remain to be effectively tackled. For example, the EPA water quality standards were violated by 13 percent of all U.S. communities, and over 100 U.S. cities regularly exceeded acceptable ozone levels because of smog buildup.

Two design related environmental issues are discussed in the following paragraphs.

12.4.1 NOISE

Noise is defined as unpleasant or unwanted sound, which means it is disturbing to or has damaging effects on wildlife, domestic animals, or humans. Even though human hearing is not as good as that of many animals, humans can hear sounds in the frequency range of approximately 15 Hertz (Hz) to 20,000 Hz, as well as sounds over a range of sound pressures of roughly 0.0002 μbar to 10,000 μbar [11, 12]. The sound pressure level in decibels (dB) is expressed as

$$P_{SL} = 10 \log_{10} \left(\frac{\rho^2}{\rho_r^2} \right) \qquad \text{[12.11]}$$

where

  ρ is the sound pressure expressed in μbar.

  ρ_r is the reference pressure (i.e., 0.0002 μbar)

The scale weighted to correct for human hearing response to varying frequencies is known as the A-weighted decibel (dBA). The sound pressures in μbar (dBA) for some common sounds, such as normal conversation, a busy intersection, a jet aircraft at 20 feet, a power lawn mower, and an automobile at 20 feet, are 0.20 (60), 6.3 (90), 2,000 (140), 20 (100), and 1 (74), respectively.

Two people approximately one meter apart having a normal conversation produce a sound pressure of roughly one-millionth of one bar [13] received by the hearer.

Sound levels dissipate with distance from its source, as expressed by the following relationship:

$$P_{SLA} = P_{SLB} - 20 \log_{10} \frac{d_A}{d_B} \qquad \text{[12.12]}$$

where

  P_{SLA} is the sound pressure level received at distance d_A from the original source.

  P_{SLB} is the sound pressure level received at distance d_B from the original source.

The sound pressure level at 120 dBA is the human pain threshold, and hearing damage can occur due to prolonged exposure to sound at 80–90 dBA. Maximum allowable exposures, expressed in hours per day, for selected noise levels (in dBA) at slow response are: 8 (90), 4 (95), 2 (100), 1 (105), and 0.5 (110).

12.4.2 AIR POLLUTION

Over time, air pollution has increased manyfold, particularly in this century. The problem of air pollution goes back centuries. For example, in 1272 AD, the consumption of coal in the city of London was banned temporarily to help clean the air [14]. Today, some of the basic major air pollutants are: carbon monoxide (CO), nitrogen oxides (NO_x), volatile organic compounds (VOC), and sulfur oxides (SO_x). Carbon monoxide is formed by the incomplete combustion of carbonaceous fuels. The major sources of CO are: aircraft, automobiles, buses, trucks, etc. In 1977, these vehicles exhausted over 85 million metric tons of carbon monoxide.

The production of NO_x is caused by fuel burned at a high temperature in air. Usually, small amounts of unburned fuel are exhausted to the air by automobiles and other such vehicles, making them the major sources of VOC. Also, substantial VOC emissions are generated through the production, refining, and marketing of petroleum.

Sulfur oxides are produced when a fuel containing sulfur is burned. Even though the major source of its production is fossil-fuel combustion, nonferrous metal smelting is also an important contributor.

To protect the public health, the Environmental Protection Agency has set national ambient air quality standards for major pollutants, as given in Table 12.4 [12].

Table 12.4 National ambient air standards for major pollutants

Major Pollutant	Standard ($\mu g/m^3$)	Average Time in Hours
SO_x	80	Yearly average
	365	24
NO_x	100	Yearly average
CO	10(mg/m^3)	8
	40(mg/m^3)	1
VOC	160	3 (6–9 AM)

Two models used to predict pollutant concentrations are described in the following paragraphs.

Box Model   The box model was developed to predict pollutant concentrations under an inversion [15]. Since the calculations associated with this model are based on a rectangular shaped area, the model is known as the box model. The box is assumed to align with wind direction, and its height is determined by the height of the inversion layer. Pollutants can come into the box through two ways, upwind ground-based sources, and wind can carry them out of the box. Under the well-mixed air condition, the steady-state outlet concentration of pollutants may be estimated from the following relationship:

$$K_p = K_{in} + \frac{E_s L_b}{S_w h} \qquad \textbf{[12.13]}$$

where

K_p is the steady-state outlet (i.e., out of the box) concentration of pollutants, expressed in $\mu g/m^3$.

K_{in} is the upwind inlet (i.e., into the box) concentration of pollutants, expressed in $\mu g/m^3$.

E_s is the emission from ground-based sources, expressed in $\mu g/m^2$–s.

L_b is the box's length in meters.

S_w is the wind speed, expressed in meters per second (m/s).

h is the height of the inversion layer above the ground, expressed in meters.

Example 12.1 | **A**ssume that the area occupied by a city resembles a box of rectangular shape, with dimensions 15 by 25 kilometers (km), and that it is covered by an inversion layer 400 meters above the ground. A wind blowing at 4 m/s parallel with the 25 km side of the city is considered clean. The emission of SO_2 from ground-based sources in the city is approximately 0.002 mg/m^2–s. Determine the steady-state SO_2 concentration in the air surrounding the city.

Solution

Substituting the given data into Equation 12.13, we get

$$K_p = 0 + \frac{(0.002)(1000)(25,000)}{4(400)}$$
$$= 31.25 \ \mu g/m^3$$

Thus, the value of the steady-state concentration of SO_2 in the air surrounding the city is 31.25 $\mu g/m^3$.

Normal Dispersion Model   In many industrial setups, such as coal-fired electricity-generating power plants, large amounts of pollution are passed through tall stacks, allowing "time and space" for the pollutants to disperse before coming back to ground level. These pollutants travel with the average wind and spread out in horizontal and vertical directions from the plume's centerline. The rate of spread in each direction is a function of such factors as pollutant characteristics, micrometeorological conditions, local geographical features, etc. This spread of pollutants may be approximated by a normal distribution [16]. The steady-state concentration of a pollutant at point (x, y, z) from an elevated source is expressed as follows:

$$K = \frac{R_e}{2\pi s \sigma_y \sigma_z} e^{\left(-\frac{1}{2}\frac{y^2}{\sigma_y^2}\right)} \left[e^{\left(-\frac{1}{2}\frac{(z-h)^2}{\sigma_z^2}\right)} + e^{\left(-\frac{1}{2}\frac{(z+h)^2}{\sigma_z^2}\right)} \right] \qquad \textbf{[12.14]}$$

where

K is the steady-state concentration of pollutants at point (x, y, z), expressed in $\mu g/m^3$.

s is the mean speed of the wind at the stack height, expressed in meters per second (m/s).

R_e is the emissions rate, expressed in $\mu g/s$.

h is the effective height of the emissions, expressed in meters, which is composed of the stack height and the height of the plume above the stack height.

z is the vertical distance (height) of the plume above the ground, expressed in meters.

y is the horizontal distance of the plume from the plume centerline, expressed in meters.

σ_z is the standard deviation of the plume concentration distribution in the vertical direction, expressed in meters.

σ_y is the standard deviation of the plume concentration distribution in the crosswind direction, expressed in meters.

x is the distance in the direction of the average wind, expressed in meters.

12.5   CASE STUDY: TACOMA NARROWS BRIDGE FAILURE

The Tacoma Narrows bridge disaster (Figure 12.1) occurred on November 7, 1940, and is regarded as one of the most spectacular bridge failures. It was the third longest bridge in the world, with a 2,800-ft main suspension span, and was built at a cost of $6,400,000 [17, 18]. On the morning of the disaster, the wind was reported to be blowing at 42 miles per hour. Subsequent analysis indicated that the failure appeared to have started at midspan with buckling of the stiffening girders, although lateral bracing may have initially parted.

A board was appointed to investigate this disaster. The board concluded in its final report that the Tacoma Narrows bridge failure resulted from excessive oscillations caused by the wind. The designers failed to take this important environmental factor adequately into account.

12.6   PROBLEMS

1. Discuss four major classifications of material.
2. Describe at least six factors to be considered in selecting material.
3. List at least eight classifications of manufacturing processes.

Figure 12.1:   Tacoma Narrows bridge

| UPI-Bettman

4. Describe the following:
    a. Forging.
    b. Casting.
    c. Machining.

5. Define the following terms:
    a. Turning.
    b. Milling.
    c. Grinding.
    d. Broaching.
    e. Drilling.

6. Describe the following types of welding:
    a. Resistance welding.
    b. Oxyfuel welding.
    c. Arc welding.
    d. Solid-state welding.

REFERENCES

1. Dieter, G.E. *Engineering Design: A Materials and Processing Approach.* New York: McGraw-Hill, 1983.

2. DeGarmo, E.P.; J.T. Black; and R.A. Kohser. *Materials and Processes in Manufacturing.* New York: Macmillan, 1988.

3. Walton, J.W. *Engineering Design: from Art to Practice.* New York: West Publishing Co., 1991.

4. Ray, M.S. *Elements of Engineering Design.* Englewood Cliffs, NJ: Prentice-Hall, 1985.

5. Ullman, D.G. *The Mechanical Design Process.* New York: McGraw-Hill, 1992.

6. Ross, B. "What Is a Design Defect?" In *Structural Failure, Product Liability and Technical Insurance.* ed. H.P. Rossmanith. New York: Elsevier Science Publishing Company, 1984, pp. 33–37.

7. *Design of Weldments.* Cleveland, OH: James F. Lincoln Arc Welding Foundation, 1963.

8. Boothroyd, G.; C. Poli; and L.E. Murch. *Automatic Assembly.* New York: Marcel Dekker, 1982.

9. Boothroyd, G.; and A.H. Redford. *Mechanical Assembly.* London: McGraw-Hill, 1968.

10. Ertas, A.; and J.C. Jones. *The Engineering Design Process.* New York: Wiley, 1993.

11. Canter, L.W. *Environmental Impact Assessment.* New York: McGraw-Hill, 1977.

12. Wanielista, M.P.; Y.A. Yousef; J.S. Taylor; and C. David Cooper. *Engineering and the Environment.* Monterey, CA: Brooks/Cole Engineering Division, 1984.

13. Mestre, V.E.; and D.C. Wooten. "Noise Impact Analysis." In *Environmental Impact Analysis Handbook.* ed. J.G. Ran, and D.C. Wooten. New York: McGraw-Hill, 1980.

14. Wark, K.; and C.F. Warner. *Air Pollution: Its origin and control.* New York: Harper and Row, 1981.

15. Wanta, R.C. "Meteorology and Air Pollution." In *Air Pollution.* ed. A.C. Stern. New York: Academic Press, 1968.

16. Turner, D.B. *Workbook of Atmospheric Dispersion Estimates.* Report No. AP-26. Washington, D.C.: Environmental Protection Agency, 1970.

17. Ross, S.S. *Construction Disasters: Design Failures, Causes, and Prevention.* New York: McGraw-Hill, 1984, pp. 216–239.

18. Watson, S.R. "Civil Engineering History Gives Valuable Lessons." *Civil Engineering,* May 1975, pp. 48–51.

chapter

13

VALUE ENGINEERING, CONFIGURATION MANAGEMENT, CONCURRENT ENGINEERING, AND REVERSE ENGINEERING

13.1 INTRODUCTION

Value engineering is primarily a function-oriented discipline applicable to cost reduction and prevention. The published literature contains various definitions for value engineering. For our purposes, we will describe value engineering as a systematic and creative technique for performing a required function at the minimum cost. In general, the two objectives of value engineering are: (a) reducing cost, and (b) improving quality through better design and manufacturing methods.

The concept of value engineering is not new. Its history goes back to 1947 when L. D. Miles of the General Electric Company first conceived the idea of value analysis. The merit of this technique was recognized by the United States Navy Bureau of Ships, and several value engineering organizations were established in the 1950s. Between 1963 and 1966, the United States Department of Defense estimated its savings to be more than $1.1 billion, due to the use of the value engineering approach [1]. Two other important milestones in the development of value engineering were: establishment of the American Society of Value Engineers in 1959, and

the first publication of the *Journal of Value Engineering* in 1962.

Configuration management is the method used to mutually identify equipment and documentation [2]. The missile launch program in the United States in the 1950s was the initial impetus for configuration management. Yet, it was not until 1962 that a formal document entitled, "Configuration Management During the Development and Acquisition Phases," AFSCM 375-1, was published by the United States Air Force (USAF). Since then, the United States military has produced several other documents on the subject:

- MIL-STD-482, "Configuration Status Accounting Data Elements and Related Features," 1970.

- MIL-STD-1456, "Contractor Configuration Management Plans," 1972.

- Directive 5010.19, "Configuration Management," 1979.

Concurrent engineering is the parallel or concurrent product design and manufacturing process development [3]. The application of concurrent or simultaneous engineering is seriously considered in a number of engineering sectors.

Reverse engineering is relatively new, beginning back in the mid-1980s. Reverse engineering is a four-stage process to develop technical data for the purpose of supporting the effective use of capital resources and enhancing productivity.

Value engineering, configuration management, concurrent engineering, and reverse engineering are discussed separately in the following sections.

13.2 VALUE ENGINEERING

Value engineering is a well-developed discipline covered by numerous publications and specific courses taught at institutions of higher learning. Before presenting the various elements of value engineering, we must first examine the reasons for poor value. The primary reason is the lack of a formal approach to achieve high value. Other factors include: resistance to change, lack of new ideas, lack of relevant information, use of temporary decisions, and reluctance to seek advice from others. The elements of good value engineering are discussed in the following sections.

13.2.1 PHASES

A value engineering study may be executed in five phases [4]: information gathering, functional analysis, speculation, evaluation, program planning and presentation, and implementation. The information phase is concerned with collecting the relevant information on such topics as: application, objective, support and testing needs, environmental and physical needs, useful life period, and potential problem areas. The functional analysis phase seeks answers to such questions as: What is its function, and what is the worth of that function? The speculation phase is basically concerned with creativity and seeks answers to such questions as, What are the possible ways to accomplish the same function? The mental processes involved here are creative and judicial.

The evaluation phase involves developing choices among the various alternatives. This is generally accomplished by performing feasibility and resource analyses. The basic objective of the program planning and presentation phase is to "sell" the value engineering proposal to concerned people. During the implementation phase, the value engineering proposal is implemented, with the approval of management, and its associated results are subsequently monitored on a regular basis.

13.2.2 VALUE ENGINEER'S TASKS

A value engineer performs a variety of tasks, some of which are as follows:

1. Select the most desirable manufacturing process.
2. Identify those quality factors that are important in increasing or decreasing the product value.

3. Select the most desirable design techniques.

4. Choose the most desirable material for the task.

5. Select the most appropriate equipment and tools to assemble the item.

6. Identify product value degradation cost factors.

To perform these tasks, a value engineer must ask a number of questions, such as those given in Table 13.1.

Table 13.1   Questions for accomplishing value engineering objectives

No.	Question
1	What is the item?
2	What is the function to be performed?
3	What will it cost?
4	What are the possible alternatives?
5	What is the cost associated with each alternative?
6	Will alternatives enhance quality?

In evaluating the market performance of a product, the value engineer uses the following guidelines:

1. Review customer/buyer reactions.

2. Review the product's performance with respect to competition, and highlight weakness causes.

3. Examine the product's sales potential.

4. Review the results of comparative analyses.

5. Review high profit margin products.

13.2.3 GUIDELINES

There are several items that could be useful in performing value engineering studies [6]:

1. Identify primary and secondary functions, and conduct cost analysis.

2. Utilize standards.

3. Obtain new information wherever possible, and then make your own decision.

4. Benefit from good human relations.

5. Identify important facts and figures

6. Use company resources effectively.

7. Take advantage of creativity approaches.

Past experience indicates that companies give various reasons for not having a value engineering program [1]:

1. Small size of the organization.
2. Service-oriented company.
3. The use of a large number of purchased parts in the manufactured product.
4. A wide variation in product size, price, use, and quality.

13.3 CONFIGURATION MANAGEMENT

During the development and operation of an engineering system many changes may occur with respect to such areas as performance, appearance, size, and weight. Configuration is the basic discipline concerned with such changes. The changes associated with a product may be classified into three categories [5]:

1. Product deficiency changes.
2. Product life-cycle cost changes.
3. Product logistical support or operational use changes.

The reasons cited for changes during the design and production phases of a product include:

1. Handling manufacturing difficulties.
2. Reducing manufacturing cost.
3. Rectifying product design deficiencies.
4. Taking advantage of the availability of better parts or components.
5. Implementing new ideas generated by the product designer.

13.3.1 TECHNIQUES

Again, configuration management is the science of organizing and controlling the planning, design, and hardware operations of a product through regular configuration control, identification, and accounting practices [7]. The techniques used in configuration management are: identification, control, and accounting and auditing. Control is a continuous function that begins at the early phase of product development and terminates at product retirement. The main purpose of configuration control is to ensure that the actions taken in each phase are appropriately approved and are adequately taken into account in the baseline documentation. Identification means the complete up-to-date description of a product during each phase. The main purpose is to ensure that product hardware and support documents are compatible during the product's entire life span. Accounting is concerned with developing an effective approach for tracking document product details and revisions, for the purpose of satisfactory retrieval. Auditing provides periodic feedbacks on the status of changes.

13.3.2 EFFECTIVE FEATURES, AND MANAGER'S QUALITIES

To determine the effectiveness of the configuration management practices, factors such as the ones listed here are considered [7]:

1. Customer satisfaction.
2. Simple identification, control, and accounting.
3. Accurate change and identification accounting.
4. Minimum manpower requirements.
5. Effective evaluation and processing of product changes.
6. Effective project team effort.
7. Complete and accurate change descriptions.
8. Minimum required documentation to meet set objectives.
9. Efficient, accurate definitions of management objectives and procedures.

Just as with general management, several managerial skills are required to lead a configuration management team. Some of these are:

1. A basic knowledge of such subjects as manufacturing, design, quality control, testing, and reliability.
2. An ability to organize and perform administration tasks effectively.
3. Tactfulness.
4. Diplomacy.
5. An ability to document effectively.
6. A knowledge of the product and of the whole organization.
7. Flexibility.

13.3.3 APPLICATIONS: RESEARCH, DESIGN AND DEVELOPMENT, MANUFACTURING, AND OPERATION

Configuration management can be useful in such situations as: poor monitoring and control of changes, and poor direction from management during the research phase. In addition, during the design and development phase, the following circumstances are prime areas for the application of configuration management:

1. Poor communication with concerned agencies.
2. Poor management leadership.
3. Poor program planning, documentation, and standardization.
4. Poorly defined design and development objectives and procedures.
5. Poorly controlled product design changes.
6. Inadequate data retrieval system.

Similarly, in the product manufacturing phase, there are several circumstances under which configuration management could be effectively applied:

1. Ineffective data storage and retrieval system.
2. Poorly described maintenance and spares procedure data.
3. Poorly controlled revisions.
4. Incompatible engineering and design specification documentation.

Factors such as the following call for the application of configuration management:

1. Poor interactions between equipment manufacturer and user.
2. Poor documentation of field modifications.
3. Poorly controlled decisions with respect to product modification.

13.4 CONCURRENT ENGINEERING

Although several organizations were practicing the concepts of concurrent engineering prior to its actual definition, the 1980s may be regarded as the real beginning of concurrent engineering, also known as simultaneous engineering [3]. This was about the time that companies in the marketplace felt the effects of such influences as more complex products, larger organizations, and newer and more innovative technologies. This scenario forced such companies to search for new product development approaches. In 1982, the United States Defense Advanced Research Projects Agency (DARPA) initiated a study to develop approaches to enhance concurrency in the product design process. This is regarded as an important milestone in the history of concurrent engineering [8]. The term concurrent engineering was coined in 1986 in Report R-338 of the Institute for Defense Analyses (IDA). Since then, several publications in the field have appeared.

13.4.1 DEFINITIONS AND COMPONENTS

Several definitions of concurrent engineering are being used in the published literature. One of those definitions is summarized as follows [9]:

Concurrent engineering is a systematic approach concept for the integrated, simultaneous design and development of both the items and their associated processes, such as manufacturing and testing. For our purposes, we will consider that concurrent engineering is composed of at least six components (activities) that go hand in hand:

1. Design for performance.
2. Manufacturability.
3. Quality.

4. Testability.

5. Serviceability.

6. Compliance.

There are several other factors that may also enter into the product design equation [9]:

1. Thermal analysis.

2. Safety.

3. Packaging design.

4. Human factors.

5. Reliability.

6. Environmental hazard analysis.

Most factors such as these are considered subsets of the first six components.

13.4.2 IMPACT AND OBJECTIVES

Concurrent engineering has impacted various areas: employees and their habits, design engineers and the tasks they performed, and managers and their leadership skills and management approaches. The objectives for the application of concurrent engineering include:

1. Reduce product development cost.

2. Reduce testing cost.

3. Reduce marketing time.

4. Reduce servicing cost.

5. Reduce manufacturing cost.

6. Improve product quality.

7. Increase profit margin.

13.4.3 GUIDELINES

There are several useful guidelines [9] for the effective application of concurrent engineering, including:

1. Form multifunctional design groups.

2. Involve suppliers and other related bodies at an early stage of the product design phase.

3. Develop and apply ways and means to improve communication with customers.

4. Design and develop processes concurrently with the design and development of the product.

5. Simulate product and process performances.

6. Integrate technical reviews into the design and development processes.

7. Incorporate the results of previous experiences.

8. Incorporate computer aided engineering (CAE) tools with the product model.

9. Continually improve the effectiveness of the design process.

Various culturally related conflicts can also arise with respect to concurrent engineering. Such conflicts may provide a healthy input to the practice of concurrent engineering. Useful guidelines for managing such conflicts in the concurrent engineering environment are as follows:

1. Investigate the real reason for the conflict.

2. Carefully evaluate each group member's input.

3. Pay careful attention to a customer's internal and external needs.

4. Generally, aim for a compromise.

5. Practice consensus management as much as possible.

13.4.4 APPLICATION AND BENEFITS

Over the years, many companies have applied concurrent engineering with considerable success [9]. For example AT&T used it to design a "circuit pack" for its 3B series computer, and IBM used it in the development of an automated electronic design automation system. Other companies reporting similar success include: Hewlett-Packard, Boeing Ballistic Systems Division, Texas Instruments, and Tektronic. One case study [3] reported several tangible benefits: product cost reduction; product quality improvement; manpower development through product experience sharing; improved procurement, process, production techniques, and quality; and total program compression. Three reported drawbacks were: incremental increase in design time, cultural changes within the organization, and the need for good risk management.

13.5 REVERSE ENGINEERING

For the smooth and continuous functioning of any production or manufacturing facility, the availability of technical data is crucial. Reverse engineering is the development of technical data critical to the support of an already manufactured item [1].

The concept of reverse engineering may not be new; it appears to have been practiced in one form or another in the past. However, according to K.A. Ingle [11], a United States Government Federal Acquisition Regulation in the mid-1980s stated, probably for the first time, that reverse engineering should be considered only when economic feasibility warrants and when all other options for developing such technical data are not promising. These data could be in the form of equipment specifications, performance characteristics, engineering drawings, etc. The data sources include: repair records, technical manuals, and performance criteria.

Today, most of the literature on reverse engineering is aimed at reverse software engineering; only a small percentage of the literature is concerned with hardware systems. The reverse engineering concept may be expressed as a four-stage process for the development of technical data to support the effective utilization of capital resources, as well as to enhance productivity.

The four stages of the reverse engineering process are as follows:

1. Data evaluation and verification.
2. Technical data generation.
3. Design verification.
4. Project implementation.

Note that all stages are conducted only after a comprehensive prescreening of the potential candidates. The data evaluation and verification stage is the most demanding element of the reverse engineering process, since more actions must be taken during this stage than in any other. The basic steps involved in this stage are:

1. Visual and dimensional inspection.
2. Discrepancy review versus available data.
3. Failure analysis.
4. Quality evaluation report generation.
5. Evaluation and verification report generation.
6. Go/no-go decision making.

The technical data to be generated include: dimensions, tolerances, quality assurance requirements, surfaces, finishes, interfaces, and performance and testing specifications. The main objective at this stage is to develop a comprehensive data package sufficient for the fabrication and acquisition of the part in the future. Some of the elements of the technical data package are: developmental design drawings; conceptual design drawings; product design drawings; commercial drawings; special inspection equipment drawings and associated documentation; specifications; special tooling drawings; preservation, packaging packing, and marking data; software and software documentation; and test requirement documents.

The objective of the design verification is to establish proof through testing that the resulting item is as originally intended.

The project implementation stage is the final stage of the reverse engineering process.

13.5.1 APPLICATION CANDIDATES

The application of reverse engineering is a business venture; therefore, the candidates for its application must be carefully chosen. Generally, a good reverse engineering candidate exhibits such characteristics as: high annual usage, high failure rate, or excessive cost. Other factors to be considered are: adequacy and availability of technical data, lack of supply, support obsolescence, or patent rights. The four characteristics that can help to identify a good candidate are: logistics, economics, technical complexity and criticality, and return on investment.

13.5.2 REVERSE ENGINEERING TEAM

The reverse engineering approach is performed by a group of people, who may individually belong to different professional categories, such as:

1. Engineers.
2. Draftsmen.
3. Technicians.
4. Shop personnel.
5. Estimators.
6. Production–manufacturing workers.

In addition, the effective application of reverse engineering requires a wide variety of specialists in such areas as: circuit design, ceramics, metallurgy, vibration analysis, etc. Even though a multitude of people form the entire engineering team, only a small number form the core reverse engineering team. The services of the remaining people are important, but not on a permanent basis. For the sake of consistency and experience, every effort is made to keep the same core team members from project to project. The person appointed to lead the core team is usually a generalist with some knowledge in such engineering disciplines as mechanical, electrical, electronic, industrial, process, and/or manufacturing. Since the core team leader interacts with people in management, other core team members, and approval agents, that leader must possess excellent communications skills, as well as other managerial qualities.

13.5.3 TRADITIONAL VERSUS REVERSE ENGINEERING DESIGN PROCESSES

A traditional design process may have from 4 to 25 stages. For example, the four stages might be: (a) need, (b) design idea, (c) prototype and test, and (d) product. On the other hand, the reverse engineering design process can have such stages as: (a) product, (b) disassembly, (c) measure and test, (d) design recovery, (e) prototype and test, and (f) reverse engineered product.

13.5.4 REVERSE ENGINEERING ADVANTAGES, RISK OF FAILURE, AND PROBABILITY OF SUCCESS

Reverse engineering has several advantages, some of which are: enhanced ability to maintain high-performance manufacturing capability, cost reduction, and useful stopgap measure to increase system productivity until required resources and other factors are available for full modernization. In essence, reverse engineering is aimed at modernizing single system components, rather than entire systems, to maintain or improve system productivity.

Reverse engineering is not risk free. Some of the potential problem areas are as follows:

1. Not all candidates for reverse engineering may be successful. Over the long term, the overall success rate of prescreened candidates could only be 65 to 75 percent.

2. The expected return on investment for prescreened candidates must be at least 25:1.

3. The expected return on investment for special projects with a high risk must be at least 200:1.

4. A low risk of failure means a high probability of success.

Strategies for increasing the success rate are as follows:

1. Invest moderately at the beginning of the reverse engineering program. Also, select only those candidates that have a high probability of success with minimal investment.

2. Expect at least a 25-percent reduction in the unit cost of an item, due to reverse engineering application.

3. Expect it to take from 2 to 5 years for a good reverse engineering program to become self-sufficient.

4. Consider reverse engineering as a tool, not the total domain of the engineer.

13.6 PROBLEMS

1. Define the following three terms:
   a.   Concurrent engineering.
   b.   Value engineering.
   c.   Configuration management.
2. What are the tasks of a value engineer?
3. What are the similarities and differences between concurrent engineering and value engineering?
4. Describe the features of effective configuration management.
5. What are the important objectives of concurrent engineering?
6. Discuss the histories of value engineering, configuration management, and concurrent engineering.
7. List the advantages and disadvantages of concurrent engineering.

REFERENCES

1. "Value Engineering." In *Engineering Design Handbook*. AMCP 706-104. Springfield, VA: National Technical Information Service, July 1971.

2. Juran, J.M.; F.M. Gryna; and R.S. Bingham. *Quality Control Handbook*. New York: McGraw-Hill, 1979.

3. Thompson, W.L. *Simultaneous Engineering: A Case Study*. Paper No. 891950. Warrendale, PA: Society of Automotive Engineers, 1989.

4. Dhillon, B.S. *Engineering Management*. Lancaster, PA: Technomic Publishing Company, 1987.

5. Dhillon, B.S. *Systems Reliability, Maintainability and Management*. New York: Petrocelli Books, 1983.

6. Heller, E.D. *Value Management: Value Engineering and Cost Reduction*. Reading, PA: Addison-Wesley, 1971.

7. Samaras, T.T.; and F.L. Czerwinski. *Fundamentals of Configuration Management*. New York: Wiley, 1971.

8. Carter, D.E.; and B.S. Baker. *Concurrent Engineering*. Reading, MA: Addison-Wesley, 1992.

9. Turino, J. *Managing Concurrent Engineering*. New York: Van Nostrand Reinhold, 1992.

10. Desa, S.; and J.M. Schmitz. "The Development and Implementation of a Comprehensive Concurrent Engineering Method: Theory and Application." *IEEE Spectrum*, 1989, pp. 1041–1049.

11. Ingle, K.A. *Reverse Engineering*. New York: McGraw-Hill, 1994.

14

ETHICAL AND LEGAL FACTORS

14.1 INTRODUCTION

The work of engineers generally affects the day-to-day life of all humans. In the pursuit of their professional tasks, engineering professionals must perform their duties in an ethical manner, to win the confidence of the public, employers, colleagues, and others. Ethics concerns what is acceptable versus what is unacceptable with respect to personal and professional obligations and moral duties. Since engineers are involved in the use of raw materials and end products, they are probably more exposed to ethical pressures than many other professionals. They may be faced with such questions as [1]:

1. Are the raw materials and end products being utilized according to acceptable ethical norms, as well as to fulfill ethical objectives?

2. Should the engineer get involved in the manufacture of a product that may be harmful to humans?

As is obvious from these questions, such analyses are not easy, and they require careful judgment on the part of the engineers.

Another important aspect of engineering design is legal factors and related issues [2–5]. The branch of law governing the liability of a product's manufacturer is known as product liability. In recent times, changes in the interpretation of the law have led to an alarming increase in product liability claims. For example, in 1976, consumer-initiated lawsuits in the United States were on the order of 50,000, and the predicted number for the year 1980 was 1,000,000 [3]. In the past, in the United States, the manufacturer of a product was only liable when negligence or unreasonable carelessness could be proved; today, however, a strict standard of liability is applied [6]. Under such conditions, the plaintiff is required to prove the following:

1. The injury caused is appropriately assignable to the defect in question.

2. The injury was caused by the defect in question.

3. The product is defective and unreasonably dangerous.

4. The defect was known at the defendant's facilities.

This chapter discusses various ethical and legal factors directly or indirectly related to engineering design.

14.2   ETHICAL CONCERNS

There are wide areas in which ethical problems may arise for engineers. Some of these are as follows:

1. Product and service advertisements.
2. Bidding for contracts.
3. Kickbacks.
4. Cartels.
5. Trade secrets.
6. Patents.

Typical ethics related questions associated with areas such as those listed include the following [1]:

• Is it ethical to accept commissions from contractors or manufacturers?

• Is it ethical to fix the price of goods produced by your company with competitors?

• Is it ethical to use some other company's trade secret to avoid bankruptcy of your company?

• Is it ethical to use your company's facilities for personal use?

• Is it ethical to design a product that directly or indirectly could be harmful to humans?

• Is it ethical to alter the test data on a product at the insistence of your employer?

14.3   ETHICS CODE

To the best of the author's knowledge, there is no single standard code of ethics document for use by engineers. This could be due to the fact that there is such a large number of engineering societies. For example, in the United States alone, there are over 150 engineering or associated societies [1].

Some of these societies have developed a code of ethics to help their members answer ethical questions that arise.

In 1912, the Institute of Electrical and Electronic Engineers (IEEE) was the first professional engineering society in the United States to adopt such a code of ethics. In summary, the four basic articles of the IEEE code are as follows:

1. **Article 1**. In this article, members are encouraged to maintain a high degree of diligence, creativity, and productivity, as well as to accept responsibility for their actions, be honest in stating claims based on available data, advance the integrity and prestige of the profession, etc.

2. **Article 2**. This article is concerned with treating colleagues and others fairly, giving proper credit to the contributions of others, helping colleagues and others in their professional development, and so on.

3. **Article 3**. This article is basically concerned with relations with clients and employers. This includes faithfulness to employers, confidentiality of technical information, acceptance of gifts, and so on.

4. **Article 4**. This article is concerned with responsibility to the public on matters such as safety, welfare, and health.

The American Society of Mechanical Engineers (ASME) has also developed its own code of ethics. The "Code of Ethics of Engineers" appeared in the ASME constitution as Article C2.1.1 and is given in Reference [1].

14.4 SELECTED ETHICS PUBLICATIONS

There are many excellent published materials on ethics, and some of the ones useful for engineers are listed here:

1. Martin, M. *Ethics in Engineering.* New York: McGraw-Hill, 1983.

2. Schaub, J.H. *Engineering Profession and Ethics.* New York: Wiley, 1983.

3. Gunn, A.S.; and P.A. Vesilind. *Environmental Ethics for Engineers.* Chelsea, MI: Lewis Publishers, 1986.

4. Andrews, K.R. *Ethics in Practice.* Cambridge, MA: Harvard Business School Press, 1989.

5. Vesilind, P.A. "Rules, Ethics, and Morals in Engineering Education." *Engineering Education,* 1988, pp. 289–293.

6. King, W.J. "The Unwritten Laws of Engineering." *Mechanical Engineering* 66, (1944), pp. 5–7.

7. Berube, B.G. "A Whistle-Blower's Perspective of Ethics in Engineering." *Engineering Education*, 1988, pp. 294–295.

8. Davenport, W.H.; and D. Rosenthal. *Engineering: Its Role and Function in Human Society.* New York: Pergamon Press, 1967.

9. Baum, R.J. *Ethics and Engineering Curricula.* New York: Institute of Society, Ethics, and the Life Sciences, 1980.

14.5 LEGAL FACTORS

In designing engineering products, design professionals must keep certain legal factors in mind. These factors may be in various forms: government legislation, patents, copyrights, liability-related court decisions, etc. Each of these items may have a severe impact on the success or failure of a product. Some of the items directly or indirectly related to legal factors are discussed separately in the following sections.

14.5.1 PRODUCT LIABILITY

Today, product liability is an important consideration in engineering design, as the cost resulting from product liability suits is increasing at an alarming rate. Engineering negligence accounts for approximately 25 percent of product litigation. Any one of the following three factors would be the basis for negligence in engineering design [6]:

1. Failure of the design to comply with accepted standards, or to specify materials of satisfactory strength.
2. Concealment of danger created by the design.
3. Failure of the manufacturer to provide required safety devices as part of the product design.

Also, the failure to provide suitable danger warning labels is another important area of negligence, which involved approximately 60 percent of the product liability cases. In the United States, under the strict liability theory, the plaintiff is only required to prove three factors: (a) the product defect caused the injury, (b) the defect existed at the time the product departed from the defendant's facility, and (c) the existing defect in the product was unreasonably dangerous. Product manufacturers in the United States may be legally liable on the basis of such factors as [3, 5]:

1. Manufacturing defects due to poor quality control and testing.
2. Sale of the packaged product in dangerous and incomplete form.
3. Poor data collection with respect to product failures and user complaints.
4. Susceptibility of the packaged product to safety-related handling damage.
5. Inadequate labelling with respect to possible danger and proper usage.
6. Susceptibility of the instructions to detachment from the packaged product prior to sale.
7. Inadequate recordkeeping with respect to product manufacturing, distribution, and sale.

To establish the grounds for product liability suits, lawyers must probe into many areas: design faults, product safety procedure accuracy, product advertisement faults, outlined testing procedures, raw materials quality, design specifications adherence, standard product manufacturing procedures, etc.

In response, product manufacturers have developed several points with which to defend against negligence charges. These points fall into five basic areas, as given in Table 14.1. The first point concerns the undetectability of the fault. Here, the manufacturers plead that no method exists to diagnose the fault. The manufacturers also point out that they did their best by giving special care to the design and manufacture of the product.

Table 14.1 Categories of points developed by product manufacturers to defend against negligence charges

No.	Point Description
1	Undetectability of the fault.
2	Negligence of the consumer.
3	Product compliance with necessary standards and regulations.
4	Statute of limitations.
5	State of the art.

In the case of consumer negligence the manufacturer takes the position that the conusmer knowingly used the faulty product. With respect to product compliance with necessary standards and regulations, the manufacturers simply plead that their products are compatible with government regulations, published standards, etc.

Questions such as, "When did the product accident occur?" and "When did the damages due to the accident manifest themselves?" are raised by the manufacturers under the statute of limitations. Finally under state of the art, manufacturers plead that they did their utmost to make the design of the product as safe as possible under the existing circumstances, and that no one else could have done better.

According to G. Dieter, [6] the following guidelines should be considered during the design process in order to minimize potential product liability problems:

1. Design the product warning labels and the user's manual as an integral part of the product design process, and use international warning symbols as much as possible. In addition, involve personnel from engineering, manufacturing, legal, marketing, etc., in developing the symbols and labels.

2. Strictly follow industry and government standards.

3. Carefully document such activities as: design, manufacturing, testing, and quality control.

4. Thoroughly test products prior to their release for sale.

5. Consider using improved quality control techniques, with the aim of reducing product liability problems.

14.5.2 COPYRIGHTS AND PATENTS

A copyright is the sole, absolute legal right to publish a tangible expression of work of a literary or artistic nature. A copyright safeguards against the unauthorized copying of such work by others. In the United States, the period covered by a copyright runs from the time of the work's creation, throughout the author's life, and 50 years thereafter. The following types of original work are covered by United States copyright laws:

1. Literary.
2. Sculptural.
3. Pictorial.
4. Graphic.

With respect to engineering design, the copyright law covers first-time written engineering specifications, as well as such items as drawings, models, and sketches. Note that under the copyright law in the United States, an idea cannot be copyrighted; only its tangible expression can be.

The largest pool of technological information in the world is probably the United States patent system with over four million U.S. patents. Furthermore, only approximately 20 percent of the technology comprised by these U.S. patents is available in published sources. In design work, this means that the oversight of U.S. patents could turn out to be quite costly. The United States patent system was started in 1790, and a patent may be awarded for a time period of up to 17 years under several categories [6, 7]: manufactured articles, machines, processes, human-made microorganisms, and matter compositions.

Three common criteria for awarding a patent are: (a) the newness or novelty of the invention, (b) the usefulness of the invention, and (c) the ingenuity of the invention, which means that its design is not obvious to people skilled in the art covered by the patent in question. Broadly speaking, a patent in good standing is the same as ownership of a piece of property, which can be sold or licensed at will. Additional points concerning patent ownership are as follows [8, 9]:

1. A patent prevents others from manufacturing, using, or selling the item covered by the patent, without express permission to do so.
2. The patent is transferable to a beneficiary in a will.
3. The patent is assignable to another person.

For current information on patents in the United States and Canada, the reader should contact the following offices:

1. Public Patent Search Facilities
   Patent and Trademark Office
   2021 Jefferson Davis Highway
   Arlington, VA 22161
   U.S.A.

2. Patent Office
   Consumer and Corporate Affairs Canada
   50 Victoria Street

Hull, Québec

Canada

14.5.3 PRODUCT WARRANTIES

Product manufacturers usually provide a written document, or warranty, guaranteeing the product's integrity. A warranty on a manufactured product spells out the responsibilities of the manufacturer in case the product is defective. According to a study reported by E. P. McGuire [10], out of 369 United States manufacturers surveyed, 99 percent had written warranties on their products, and the average cost of a warranty claim was less than the 2 percent of sales. The manufacturers cited the following basic reasons for providing warranties:

1. Enhancing market competitiveness.
2. Outlining responsibilities and liabilities.
3. Supporting their manufactured products.

Specific reasons for having a warranty could be as follows [11]:

1. Enhances the manufactured product's marketability and acceptance.
2. Encourages the production of better goods, as well as better monitoring of product performance.
3. Assures the customers that the purchased item will at least meet contractual specifications, and makes the product manufacturer responsible for repairing defects.
4. Serves as a means of assuring that technical services will be available as needed.
5. Serves as a useful tool for speeding up the payment for a sold item and, during the no-charge warranty timeframe, provides the buyer time to gain operating experience with the purchased goods.

According to the surveys reported in Reference 10, the most frequent reasons cited for an increase in warranty claims were: increase in buyer awareness of manufacturer's warranty obligations, misuse of product, product complexity, poor customer maintenance, product quality deterioration, and inflationary pressures. Furthermore, the servicing executives of manufacturers forecasted several future warranty related problems: increased customer expectations, government legislation, rising product liability, rapid cost increases associated with labor, material, etc., increased record-keeping requirements, such as for warranty starting dates, and so on.

A mathematical model is proposed in Reference 13 to estimate a manufacturer's warranty cost. The warranty cost is expressed as

$$WC = FC + \lambda(TOH)(ACR) \qquad \textbf{[14.1]}$$

where

WC is the manufacturer's warranty cost.

FC   is the warranty fixed cost to the manufacturer.

λ   is the constant failure rate of the warranted product, expressed in failures per hour.

ACR   is the warranted product's average cost to the manufacturer.

TOH   is the warranted product's total operating hours over the warranty period.

14.6 PROBLEMS

1. Discuss the areas in which ethics problems may arise for engineers.
2. Discuss the similarities and differences between the Institute of Electrical and Electronic Engineers (IEEE) and the American Society of Mechanical Engineers (ASME) code of ethics.
3. What is product liability?
4. Discuss the factors that are regarded as the basis for negligence in engineering design in the United States.
5. Describe the features of a copyright.
6. What is a patent?
7. What are the reasons for providing a warranty?

REFERENCES

1. Walton, J.W. *Engineering Design.* New York: West Publishing Company, 1991.
2. Weinstein, A.D.; A.D. Twerski: H.R. Piehler; and W.A. Donaher. *Product Liability and the Reasonably Safe Product.* New York: Wiley & Sons, 1978.
3. Dhillon, B.S. *Engineering Management.* Lancaster, PA: Technomic Publishing Company, 1987, Chapter 19.
4. Klein, S.J. *How to Avoid Product Liability.* Englewood Cliffs, NJ: Institute for Business Planning, 1980.
5. Kolb, J.; and S.S. Ross. *Product Safety and Liability — A Desk Reference.* New York: McGraw-Hill, 1980.
6. Dieter, G. *Engineering Design.* New York: McGraw-Hill, 1983.
7. Ray, M.S. *Elements of Engineering Design: An Integrated Approach.* Englewood Cliffs, NJ: Prentice-Hall, 1985.
8. Pressman, D. *Patent It Yourself? How to Protect, Patent, and Market Your Inventions.* New York: McGraw-Hill, 1979.

9. Burge, D.A. *Patent and Trademark Tactics and Practices*. New York: Wiley, 1979.

10. McGuire, E.P. *Industrial Product Warranties: Policies and Practices*. New York: Conference Board, Inc., 1980.

11. Flottman, W.W.; and M.R. Worstell, "Mutual Development Application, and Control of Supplier Warranties." American Airlines and Litton-Aero Products Division. *Proceedings of the Annual Reliability and Maintainability Symposium*, 1977, pp. 213–221.

12. Bonner, W.J. "Warranty Contract Impact on Product Liability." *Proceedings of the Annual Reliability and Maintainability Symposium*, 1977, pp. 261–263.

13. Balaban, H.S.; and M.A. Meth. "Contractor Risk Associated with Reliability Improvement Warranty." *Proceedings of the Annual Reliability and Maintainability Symposium*, 1978, pp. 123–129.

APPENDIX

BIBLIOGRAPHY: ENGINEERING DESIGN LITERATURE

A.1 INTRODUCTION

Over the years, many excellent publications on engineering design and related areas have appeared, in the form of textbooks, journal publications, conference proceedings papers, technical reports, and so on. This section lists selected publications, with an emphasis on textbooks. The period covered by the listing is the late 1950s to the 1990s. The purpose of this listing is to provide readers with sources of additional information in their areas of interest.

A.2 SELECTED PUBLICATIONS

1. Acret, J. *Architects and Engineers: Their Professional Responsibilities*. New York: McGraw-Hill, 1977.

2. Akagi, S.; and K. Fujita. "Knowledge Based Geometric Modelling System for Preliminary Design Using Object-Oriented Approach." *Proceedings of the 15th Design Automation Conference.* New York: American Society of Mechanical Engineers, 1989, pp. 129–134.

3. Akin, J.E. *Computer Assisted Mechanical Design.* Englewood Cliffs, NJ: Prentice-Hall, 1990.

4. Akman, V. *Knowledge Engineering in Design.* Report. Amsterdam: Center for Mathematics and Computer Science, 1987.

5. Alger, R.M.; and C.V. Hays. *Creative Synthesis in Design.* Englewood Cliffs, NJ: Prentice-Hall, 1964.

6. Altamuro, V. "Design for Automated Manufacturing." *Design Graphics World.* March 1987, pp. 15–18.

7. *An Instructional Aid for Occupational Safety and Health in Mechanical Engineering Design.* Report. New York: American Society of Mechanical Engineers, 1984.

8. Andreasen, M.M. *Design for Assembly.* New York: Springer-Verlag, 1985.

9. *Ethics in Practice.* ed. K.R. Andrews. Cambridge, MA: Harvard Business School Press, 1989.

10. Ang, A.H.S.; and W.H. Tang. *Probability Concepts in Engineering Planning and Design.* New York: Wiley, 1984.

11. Araoz, A. *Consulting and Engineering Design in Developing Countries.* Ottawa, Canada: International Development Research Centre, 1981.

12. Arora, J.S. *Introduction to Optimum Design.* New York: McGraw-Hill, 1989.

13. "Artificial Intelligence in Engineering Design: Special Issue." *The International Journal for Artificial Intelligence in Engineering,* 2, no. 3, 1987.

14. Artobolevsky, I. *Mechanisms in Modern Engineering Design.* Moscow: Mir Publishers, 1975.

15. Asch, P. *Consumer Safety Regulation.* New York: Oxford University Press, 1988.

16. Asimow, M. *Introduction to Design.* Englewood Cliffs, NJ: Prentice-Hall, 1962.

17. Bailey, R.W. *Disciplined Creativity for Engineers.* Ann Arbor, MI: Ann Arbor Science, 1978.

18. Bailey, R.W. *Human Performance Engineering: Using Human Factors/Ergonomics to Achieve Computer Systems Usability.* Englewood Cliffs, NJ: Prentice-Hall, 1989.

19. Baird, D.C. *Experimentation: An Introduction to Measurement Theory and Experimental Design.* Englewood Cliffs, NJ: Prentice-Hall, 1962.

20. Bajaria, H.J. "Motivating Design Engineers for Reliability, Part I." *Proceedings of the American Society for Quality Control Conference,* 1979, pp. 767–773.

21. Bajaria, H.J. "Motivating Design Engineers for Reliability, Part II." *Proceedings of the American Society for Quality Control Conference,* 1980, pp. 168–176.

22. Ballast, D.K. *The Expert Witness in Architectural and Engineering Design Litigation.* Monticello, IL: Vance Bibliographies, 1988.

23. Barker, B. *Quality by Experimental Design.* New York: Marcel Dekker, 1985.

24. Beakley, G.C.; and E.G. Chilton. *Introduction to Engineering Design and Graphics.* New York: Macmillan, 1973.

25. Beakley G.C.; and G.C. Ernest. *Design Serving the Needs of Man.* New York: Macmillan, 1974.

26. Beard, P.D.; and T.F. Talbot. "What Determines if a Design Is Safe?" *Proceedings of the American Society of Mechanical Engineers (ASME) Winter Annual Conference.* New York: ASME, 1990, pp. 90–WA/DE–20.

27. Besant, C.B. *Computer Aided Design and Manufacture.* New York: Wiley, 1980.

28. Black, I.; and J.D. Cross. "Mechanical Engineering Design With Computer Aided Technology." *Proceedings of the Institution of Mechanical Engineers, Part B., Management and Engineering Manufacture* 204, no. 1 (1990), pp. 29–36.

29. Blake, A. *Practical Stress Analysis in Engineering Design.* New York: Marcel Dekker, 1982.

30. Borg, S.F. *Earthquake Engineering: Mechanism, Damage Assessment and Structural Design.* Singapore: World Scientific, 1988.

31. Bralla, J.S. *Handbook of Product Design for Manufacturing: A Practical Guide to Low-Cost Production.* New York: McGraw-Hill, 1986.

32. Brandt, R. *Engineering Design Outputs.* Paper No. 81 – 495. Dearborn, MI: Society of Manufacturing Engineers, 1981.

33. Bronikowski, R.J. *Managing the Engineering Design Function.* New York: Van Nostrand Reinhold, 1986.

34. Burgess, J.A. *Design Assurance for Engineers and Managers.* New York: Marcel Dekker, 1984.

35. Burgess, J.H. *Designing for Humans: The Human Factor in Engineering.* Princeton, NJ: Petrocelli Books, 1986.

36. Burr, A.C. *Mechanical Analysis and Design.* New York: Elsevier Science, 1962.

37. Buxton, I.L. *Engineering Economics and Ship Design.* Wallsend, U.K.: British Maritime Technology, Ltd., 1987.

38. Calpan, R. *By Design.* New York: St. Martin's Press, 1982.

39. Caroll, J.T.; and T.F. Bellinger. "Designing Reliability Into Rubber and Plastic AC Motor Control Equipment." *IEEE Transactions on Industry and General Applications,* 1969, pp. 455–464.

40. Cheremisinoff, N.P. *Product Design and Testing of Polymeric Materials.* New York: Marcel Dekker, 1990.

41. Chisholm, A.W.J. "Engineering Design Analogy for Engineering Education." *Computers in Industry* 14 (1990), pp. 197–204.

42. Chow, W.C. *Cost Reduction in Product Design.* New York: Van Nostrand Reinhold, 1978.

43. Clinch, R.W. "Affective Engineering Design Education." *Proceedings of the Conference on Frontiers in Education.* New York: Institute of Electrical and Electronic Engineers, 1989, pp. 206–212.

44. Clothier, W. "Design Models Are Naturals for Managing Information." *Design Graphics World,* May 1987, pp. 40–41.

45. Clothier, W. "Good Models Never Die." *Design Graphics World,* December 1986, pp. 28–30.

46. Collins, J.A. *A Failure of Materials in Mechanical Design.* New York: Wiley, 1981.

47. Constance, J.D. *How to Become a Professional Engineer.* New York: McGraw-Hill, 1978.

48. Cook, N.H. *Mechanics and Materials for Design.* New York: McGraw-Hill 1984.

49. Cornish, E.H. *Materials and the Designer.* New York: Cambridge University Press, 1987.

50. Cross, N. *Engineering Design Methods.* New York: Wiley, 1989.

51. Crouse, R.L. "Graphic Trees Help Study of Reliability Versus Cost." *Product Engineering* 38 (1967), pp. 48–49.

52. Cullum, R.D. *Handbook of Engineering Design.* London: Butterworth, 1988.

53. Damon, A.; H.W. Stondt; and R.A. McFarland. *The Human Body in Equipment Design.* Cambridge, MA: Harvard University Press, 1966.

54. Dandy, G.C.; and R.F. Warner. *Planning and Design of Engineering Systems.* Boston: Unwin Hyman, 1989.

55. Davenport, W.H.; and D. Rosenthal. *Engineering: The Role and Function in Human Society.* New York: Pergamon Press, 1967.

56. Dean, B. "A Guide to the Management of Design." *Institute of Production Engineers Journal,* 1989, pp. 19–20.

57. Dehnad, K. *Quality Control, Robust Design, and the Taguchi Method.* Pacific Grove, CA: Wadsworth & Brooks/Cole, 1989.

58. DeJong, P.S.; J.S. Rising; and M.W. Almfeldt. *Engineering Graphics: Communication, Analysis, Creative Design.* Dubuque, IA: Kendall/Hunt, 1983.

59. *Design Handbook: Engineering Guide to Spring Design.* Bristol, CT: Associated Spring, Barnes Group, 1987

60. Deutschman, A.D. *Machine Design: Theory and Practice.* New York: Macmillan, 1975.

61. Dhillon, B.S. *Human Reliability.* New York: Pergamon Press, 1986.

62. Dhillon, B.S. *Quality Control, Reliability, and Engineering Design.* New York: Marcel Dekker, 1985.

63. Dhillon, B.S. *Reliability Engineering in Systems Design and Operation.* New York: Van Nostrand Reinhold, 1983.

64. Dickey, D.G. "Developing Your Commercial Reliability Program." *Proceedings of the American Society for Quality Control Conference,* 1980, pp. 724–728.

65. Diesch, K.H. *Analytical Methods in Project Management.* Ames, IA: Iowa State University, 1987.

66. Dieter, G.E. *Engineering Design: A Materials and Processing Approach.* New York: McGraw-Hill 1983.

67. *Dimensioning and Tolerancing Standard.* ANSIY14.5M-1982, New York: American Society of Mechanical Engineers, 1982.

68. *Directory of Short Courses for Engineering Designers.* London: Design Council, 1987.

69. Doyle, L.E. *Manufacturing Processes and Materials for Engineers.* Englewood Cliffs, NJ: Prentice-Hall, 1985.

70. Dreyfuss, H. *The Measure of Man: Human Factors in Design.* New York: Whitney Library of Design, 1967.

71. Dudley, D.W. *Handbook of Practical Gear Design.* New York. McGraw-Hill, 1984.

72. Dunn, R.M.; and B. Herzog. *Computer-Aided Design, Engineering and Drafting.* Pennsauken, NJ: Auerbach Publishers, 1984.

73. Dym, J.B. *Product Design with Plastics: A Practical Manual.* New York: Industrial Press, 1983.

74. Dyson, B.F.; and D.R. Hayhurst. *Materials and Engineering Design: The Next Decade.* Brookfield, VT: Institute of Metals, London, 1989.

75. Earle, J.H. *Engineering Design Graphics.* Reading, MA: Addison-Wesley, 1987.

76. Eder, W.E. "Design Science: Meta-Science to Engineering Design." *Proceedings of the American Society of Mechanical Engineers 2nd International Conference on Design Theory and Methodology,* 1990, pp. 327–335.

77. Eder, W.E. "Engineering Design: A Perspective on U.K. and Swiss Developments." *Proceedings of the 2nd American Society of Mechanical Engineers International Conference on Design Theory and Methodology,* 1990, pp. 225–234.

78. Eder, W.E. "Teachable Fundamentals of Engineering Design." *Proceedings of the Conference on Frontiers in Education.* New York: Institute of Electrical and Electronic Engineers, 1989, pp. 206–212.

79. Encarnacao, J.L. *Computer Aided Design: Modelling, Systems Engineering, CAD-Systems.* New York: Springer-Verlag, 1980.

80. *Engineering Design Handbook: Maintainability Engineering Theory and Practice.* AMCP 706-133. U.S. Army Material Command, 5001 Eisenhower Avenue, Alexandria, Virginia 22333, 1976, p. 5.21.

81. *Engineering Design Handbook:Development Guide for Reliability, Part II Design for Reliability.* AMCP 706-196. U.S. Army Material Command, 5001 Eisenhower Avenue, Alexandria, Virginia 22333, 1976, pp. 1.8–1.9.

82. *Engineering Design in Wood: Working Stress Design.* Report. Rexdale, Ontario: Standards Council of Canada, 1984.

83. *Engineering Drawing Practices Standard* DOD-STD-100. Washington, D.C.: Department of Defense, 1961.

84. "Engineering Education: Integrating Design Throughout the Curriculum (Special Issue)." *Journal American Society for Engineering Education* 80, no. 5 (1990), pp. 516–570.

85. Eschenauer, H.; J. Koski; and A. Osyczka. *Multicriteria Design Optimization: Procedures and Applications.* New York: Springer-Verlag, 1990.

86. *Evaluation of Design Effectiveness.* Report. Austin, TX: Construction Industry Institute, 1986.

87. *Changing Design.* ed. P.T. Evans, New York: Wiley, 1982.

88. Farr, M. *Design Management.* London: Cambridge University Press, 1955.

89. Faupel, J.; and E.F. Franklin. *Engineering Design: A Synthesis of Stress Analysis & Materials Engineering.* New York: Wiley, 1981.

90. Finkelstein, W.; and J.A.R. Guertin. *Integrated Logistics Support: The Design Engineering Link.* Kempston, Bedford, U.K.: IFS Publications, 1988.

91. Florman, S. *The Civilized Engineer.* New York: St. Martin's Press, 1987.

92. Flurscheim, C.H. *Industrial Design in Engineering.* Berlin: Springer-Verlag, 1983.

93. Foulds, L.R. *Optimization Techniques.* New York: Springer-Verlag, 1981.

94. Frazer, J.R. *Applied Linear Programming.* Englewood Cliffs, NJ: Prentice-Hall, 1968.

95. Frederick, S.W. "Human Energy in Manual Lifting." *Modern Materials Handling,* 14 (1959), pp. 74–76.

96. French, M.J. *Conceptual Design for Engineers.* London: Design Council, 1985.

97. French, M.J. *Invention and Evolution: Design in Nature and Engineering.* New York: Cambridge University Press, 1988.

98. French, T.E.; C.J. Vierck, and R.J. Foster. *Graphic Science and Design.* New York: McGraw-Hill, 1984.

99. Fry, J. "Design Management." Journal of the Royal Society of Arts, 5333 (CXXXII), (1986), pp. 304–308.

100. Fu, J.F.; R.G. Fenton; and W.L. Cleghorn. "Nonlinear Mixed Integer-Discrete-Continuous Programming and Its Application to Engineering Design Problems." *Proceedings of the 15th Design Automation Conference.* American Society of Mechanical Engineers, 1989, pp. 59–65.

101. Furman, T.T. *Approximate Methods in Engineering Design.* London: Academic Press, 1980.

102. Gerdeen, J.W. "Future of Engineering Design Management." *Proceedings of the Engineering Database Management,* 1989, pp. 39–42.

103. Gerdeen, J.W.; and P.E. Hansen. "Engineering Design Management." *Proceedings of the American Society of Mechanical Engineers International Computers in Engineering Conference and Exposition,* 1990, pp. 31–35.

104. Gero, J.S. *Design Optimization.* Orlando, FL: Academic Press, 1985.

105. Glegg, G.L. *The Design of Design.* New York: Cambridge University Press, 1969.

106. Glegg, G.L. *The Development of Design.* New York: Cambridge University Press, 1981.

107. Grandjean, E. *Fitting the Task to the Man: An Ergonomic Approach.* London: Taylor and Francis, 1980.

108. Grant, E.L.; and R.S. Leavenworth. *Statistical Quality Control.* New York: McGraw-Hill, 1980.

109. Greenwood, D.C. *Engineering Data for Product Design.* New York: McGraw-Hill, 1961.

110. Greenwood, D.C. *Product Engineering Design Manual.* Malabar, FL: Kreiger, 1982.

111. Grethers, W.F.; and C.A. Baker. *Visual Presentation of Information in Human Engineering Guide to Equipment Design.* ed. H.P. Van Cott, and R.G. Kinkade. Washington, D.C.: U.S. Government Printing Office, 1972, Chap. 3.

112. Guyer, E.C.; and D.L. Brownell. *Handbook of Applied Thermal Design.* New York: McGraw-Hill, 1989.

113. Hajek, V. *Management of Engineering Projects,* New York: McGraw-Hill, 1984.

114. Hales, C. *Analysis of the Engineering Design Process in an Industrial Context.* Eastleigh, Hants., U.K.: Grants Hill Publications, 1991.

115. Hales, C. "Analysis of Engineering Design: The Space Shuttle Challenger." *Proceedings of the International Workshop on Engineering Design and Manufacturing Management,* University of Melbourne, Australia, November 1988, pp. 89–97.

116. Hall, D.M.; P.F. Kelly; and J.R. Chandler. "Integrating Engineering Design into a Computer Aided Engineering Degree Course." *International Journal of Mechanical Engineering Education* 18, no. 1 (1990), pp. 43–52.

117. Hangen, E.B. *Probabilistic Mechanical Design.* New York: Wiley, 1980.

118. Harrisberger, L. *Engineermanship: The Doing of Engineering Design.* Monterey, CA: Brooks/Cole, 1982.

119. Hawkes, B.; and R. Abinett. *The Engineering Design Process.* London: Pitman Publishing, 1984.

120. Hayes, R.H.; and J. Ramchandran. "Manufacturing Crisis: New Technologies, Obsolete Organizations." *Harvard Business Review,* Sept./Oct. 1988, pp. 77–85.

121. Hill, P.H. *The Science of Engineering Design.* New York: Holt, Reinhart and Winston, 1970.

122. Hilton, J.R. *Design Engineering Project Management: A Reference.* Lancaster, PA: Technomic Publishing Co., 1985.

123. Hindhede, A. *Machine Design Fundamentals.* New York: Wiley, 1983.

124. Hoeppner, D.W.; and F.L. Gates. *Fretting Fatigue Considerations in Engineering Design.* International Council on Aeronautical Fatigue, 1981, pp. 155–164.

125. Hollins, W.; and S. Pugh. *Successful Product Design-What to Do and When.* London: Butterworths, 1990.

126. Horgan, J. *Engineering Design Services in an Enterprise.* Dearborn, MI: Paper No. 90-766. Society of Manufacturing Engineers, 1990.

127. Hubka, V.; M.M. Andreasen; W.E. Eder; and P.J. Hills. *Practical Studies in Systematic Design.* London: Butterworths, 1988.

128. Hubka, V.; and W.E. Eder. *Principles of Engineering Design.* London: Butterworth Scientific, 1982.

129. Hubka, V.; and W.E. Eder. *Theory of Technical Systems: A Total Concept Theory for Engineering Design.* Berlin: Springer-Verlag, 1988.

130. Huchingson, R.D. *New Horizons for Human Factors in Design.* New York: McGraw-Hill, 1981.

131. Husslage, R. "Putting Conceptual Design to Work." *Computer-Aided Engineering* 8, no. 10 (1989), p. 3.

132. Hutchingson, R.D. *New Horizons for Human Factors in Design.* New York: McGraw-Hill, 1981.

133. *Improving Engineering Design.* Report. Committee on Engineering Design Theory and Methodology, National Research Council, Washington, D.C.: National Academy Press, 1991.

134. Ishii, K.; C.H. Lee; and R.A. Miller. "Methods for Process Selection in Design." *Proceedings of the 2nd American Society of Mechanical Engineers International Conference on Design Theory and Methodology,* 1990, pp. 105–112.

135. Ivergard, T. *Handbook of Control Room Design and Ergonomics.* New York: Taylor and Francis, 1989.

136. "Special Issue on Systems Design Engineering." ed. L.A. Jackson. *British Telecom. Technology Journal* 4, no. 3, 1986.

137. Jacobs, R.M. "Design Review: A Liability Preventer." *Mechanical Engineering* 97 (1975), pp. 34–39.

138. Jakiela, M.J.; and L. Fayad. "Identification of Factors that Contribute to Engineering Design Skill." *Proceedings of the IEEE International Conference on Systems, Man and Cybernetics,* 1989, pp. 300–302.

139. Jensen, C.H. *Engineering Drawing and Design.* Toronto: McGraw-Hill Ryerson, 1980.

140. Jensen, C.H.; and J.D. Helsel. *Engineering Drawings and Design.* New York: McGraw-Hill, 1985.

141. Johnson, R.C. *Optimum Design of Mechanical Elements.* New York: Wiley, 1961.

142. Johnston, G.H. *Permafrost: Engineering Design and Construction.* New York: Wiley, 1981.

143. Jones, J.H.C. *Design Methods: Seeds of Human Futures.* New York: Wiley, 1981.

144. Jones, J.V. *Engineering Design: Reliability Maintainability and Testability.* Blue Ridge Summit, PA: TAB Professional and Reference Books, 1988.

145. *Quality Control Handbook.* ed. J.M. Juran; F.M. Gryna; and R.S. Bingham. New York: McGraw-Hill, 1979, pp. 8.17–8.21, 8.64–8.65.

146. Karr, L.J. *Designing Cost-Efficient Mechanisms: Minimum Constraint Design, Designing with Commercial Components, and Topics in Design Engineering.* New York: McGraw-Hill, 1990.

147. Katz, R.H. *Information Management for Engineering Design.* Berlin: Springer-Verlag, 1985.

148. Khan, M.; and D.G. Smith. "Overcoming Conceptual Barriers by Systematic Design." *Proceedings of the ICED.* London: Institution of Mechanical Engineers Publications, 1989, pp. 608–619.

149. Khanutin, M.B.; and G.N. Dobrin. "System for Organizing the Design Process." *Soviet Energy Technology* no. 3 (1989), pp. 66–70.

150. King, W.J. "The Unwritten Laws of Engineering." *Mechanical Engineering* 66, no. 7, 1944.

151. Krick, E.V. *An Introduction to Engineering and Engineering Design.* New York: Wiley, 1969.

152. Kumar, A. *Chemical Process Synthesis and Engineering Design.* New Delhi: McGraw-Hill, 1982.

153. Laak, R. *Wastewater Engineering Design for Unsewered Areas.* Ann Arbor, MI: Ann Arbor Science Publishers, 1980.

154. Lange, J.C. *Solving Mechanical Design Problems with Computer Graphics.* New York: Marcel Dekker, 1986.

155. Leech, D.J. *Management of Engineering Design.* New York: Wiley, 1972.

156. Leech, D.J.; and B.T. Turner. *Engineering Design for Profit.* New York: Wiley, 1985.

157. Lees, W.A. *Adhesives in Engineering Design.* London: Design Council, 1984.

158. Lentz, W. *Design of Automatic Machinery.* New York: Van Nostrand Reinhold, 1985.

159. Levary, R.R. *Engineering Design: Better Results Through Operations Research Methods.* New York: North-Holland, 1988.

160. Levens, A.S.; and W. Chalk. *Graphics in Engineering Design.* New York: Wiley, 1980.

161. Levy, S.; and J.H. DuBois. *Plastics Product Design Engineering Handbook.* New York: Chapman and Hall, 1984.

162. Lewis, W.P.; and A. Samuel. *Fundamentals of Engineering Design.* Englewood Cliffs, NJ: Prentice-Hall, 1989.

163. Love, F. *Planning and Creating Successful Engineering Designs: Managing the Design Process.* Los Angeles, CA: Advanced Professional Development Inc., 1986.

164. Love, S.F. *Planning and Creating Successful Engineered Designs.* New York: Van Nostrand Reinhold, 1980.

165. Love, S.F. "Design Methodology: Defining the Problem." *Design Engineering,* (Toronto), July 1969, pp. 69–72.

166. Lupton, T. *Human Factors: Man, Machine, and New Technology.* Berlin: Springer-Verlag, 1986.

167. Lutz, J.D.; D.E. Hancher; and E.W. East. "Framework for Design-Quality-Review Data-Base System." *Journal for Management in Engineering* 6 (1990), pp. 296–312.

168. Luzadder, W.J.; and K.E. Botkin. *Problems in Engineering Drawing for Design, Product Development, and Numerical Control.* Englewood Cliffs, NJ: Prentice-Hall, 1981.

169. Luzadder, W.J.; and J.M. Duff. *Introduction to Engineering Drawing: The Foundations of Engineering Design and Computer-Aided Drafting.* Englewood Cliffs, NJ: Prentice-Hall, 1989.

170. Luzadder, W.J. *Fundamentals of Engineering Drawing for Design, Product Development, and Numerical Control.* Englewood Cliffs, NJ: Prentice-Hall, 1981.

171. Maher, M.L. "Process Models for Design Synthesis." *AI Magazine* 11 (1990), pp. 49–58.

172. Mahmoud, M.A.M.; and S. Pugh. "The Costing of Turned Components at the Design Stage." *Proceedings of the Information for Designers Conference,* 1979, pp. 37–42.

173. Majunder, D.; R.E. Fulton; and J.Shilling. "Alternate Decomposition Hierarchy Management for Engineering Design." *Proceedings of the American Society of Mechanical Engineers International Computers in Engineering Conference and Exposition,* 1990, pp. 75–83.

174. Martin, M. *Ethics in Engineering.* New York: McGraw-Hill, 1983.

175. McCormick, E.J.; and M.S. Sanders. *Human Factors in Engineering and Design.* New York: McGraw-Hill, 1982.

176. McCuen, R.H. "Guidance for Engineering-Design-Class Lectures on Ethics." *Journal of Professional Issues in Engineering* 116 (1990), pp. 251–257.

177. Meadow, C.T. *Applied Data Management.* New York: Wiley, 1976.

178. Medland, A.J. *The Computer-Based Design Process.* London: Kogan Page, Ltd., 1986.

179. Meguid, S.A. *Integrated Computer-Aided Design of Mathematical Systems.* New York: Elsevier Applied Science Publishers, 1987.

180. Meister, D. *Conceptual Aspects of Human Factors.* Baltimore: John Hopkins University Press, 1989.

181. Meredith, D.D. *Design and Planning of Engineering Systems.* Englewood Cliffs, NJ: Prentice-Hall, 1985.

182. Merkley, C.D. *Design for Manufacturability: Organization and Engineering Problem.* Paper No. 89-478. Dearborn, MI: Society of Manufacturing Engineers, 1989.

183. Merritt, F.S.; and J.E. Ambrose. *Building Engineering and Systems Design.* New York: Van Nostrand Reinhold, 1990.

184. Messina, S. *Statistical Quality Control for Manufacturing Managers.* New York: Wiley, 1987.

185. Michaels, J.V.; and W.P. Wood. *Design to Cost.*, New York: Wiley 1989.

186. Middendorf, W.H. *Engineering Design.* Boston: Allyn and Bacon, 1969.

187. Middendorf, W.H. *Design of Devices and Systems.* New York: Marcel Dekker, 1986.

188. *Human Factors Engineer Design for Army Material.* MIL-HDBK-759A. Washington, D.C.: Department of Defense, 1968.

189. Miller, T.G. *Living in the Environment.* Belmont, CA: Wadsworth Publishing Company, 1988.

190. Mischke, C.R. *Mathematical Model Building.* Ames, IA: Iowa State University Press, 1980.

191. Mochel, E.V.; and L.S. Fletcher. *Problems in Engineering Design Graphics.* Englewood Cliffs, NJ: Prentice-Hall, 1981.

192. Montgomery, D.C. "Experimental Design and Product and Process Development." *Manufacturing Engineering,* September 1988, pp. 57–63.

193. Morris, G. *Engineering–A Design Making Process.* Boston: Houghton Mifflin, 1977.

194. Mucci, P.; and A.M. Thaine. *Handbook for Engineering Design Using Standard Materials and Components.* Durley, Southampton, U.K: P.E.R. Mucci Ltd.,1990.

195. Murrell, K.F.H. *Human Performance in Industry.* New York: Reinhold, 1965.

196. Muster, D.; and F. Mistree. "Engineering Design as It Moves from an Art Towards a Science: Its Impact on the Education Process." *International Journal of Applied Engineering Education* 5, no. 2 (1989), pp. 239–246.

197. Napier, M.A. "Design Review-As a Measure of Achieving Quality in Design." *Engineering* 219 (1979), pp. 286–289.

198. National Research Council Committee on the CAD/CAM Interface. *Computer Integration of Engineering Design and Production.* Washington, D.C.: National Academy Press, 1984.

199. Nevins, J.L.; and D.E. Whitney. *Concurrent Design of Products and Processes.* New York: McGraw-Hill, 1989.

200. Norman, D.A.; and S.W. Draper. *User Centered System Design: New Perspectives on Human-Computer Interaction.* Hillsdale, NJ: Erlbaum Associates, 1986.

201. Oborne, D.J. *Ergonomics at Work.* New York: Wiley, 1982.

202. Orpwood, R.D. "Design Methodology for Aids for the Disabled." *Journal of Medical Engineering and Technology* 14, (1990), pp. 2–10.

203. Osborn, A.F. *Applied Imagination.* New York: Scribner, 1979.

204. Owen, D.G. "Engineering Against Products Liability, Principles and Problems." *Proceedings of the Third International Conference on Structural Failure, Product Liability and Technical Insurance,* 1990, pp. 39–46.

205. Pahl, G.; W. Beitz; and K. Wallace. *Engineering Design.* London: Design Council, 1984.

206. Pahl, G. *Engineering Design: A Systematic Approach.* London: Design Council, 1988.

207. Papalambros, P.; and N. Wilde. *Principles of Optimal Design: Modelling and Computation.* New York: Cambridge University Press, 1988.

208. Papanek, V. *Design for Human Scale.* New York: Van Nostrand Reinhold, 1983.

209. Parker, M. *Manual of British Standards in Engineering Drawing and Design.* London: British Standards Institution, 1984.

210. Parker, S.P. *McGraw-Hill Dictionary of Mechanical and Design Engineering.* New York: McGraw-Hill, 1984.

211. Pellini, W.S. *Manual of Engineering Procedures for Fracture-Safe Design.* Chicago: Association of American Railroads, 1980.

212. Peters, G.A.; and B.B. Adams. "Three Criteria for Readable Panel Markings." *Product Eng.* 30, (1959), pp. 55–57.

213. Petroski, H. *To Engineer Is Human: The Role of Failure in Successful Design.* New York: St. Martin's Press, 1985.

214. Phadke, M.S. *Quality Engineering Using Robust Design.* Englewood Cliffs, NJ: Prentice-Hall, 1989.

215. Pheasant, S.T. *Bodyspace: Anthropometry, Ergonomics and Design.* London: Taylor and Francis, 1986.

216. Piatak, M.A.; and G.P. Kimball. "Automating the Design Process." *Proceedings of the American Society of Mechanical Engineers (ASME) 2nd Conference in Flexible Assembly Systems.* New York: 1990.

217. Pitts, G. *Techniques in Engineering Design.* London: Butterworths, 1973.

218. Powrie, S. "Intelligent Aids to Design: Some Interface Issues." *Institution of Electrical Engineers Colloquium on Artificial Intelligence in the User Interface.* London: 1990.

219. Priest, J.W. *Engineering Design for Producibility and Reliability.* New York: Marcel Dekker, 1988.

220. *Proceedings of the 1st International Conference on Design Theory and Methodology.* New York: American Society of Mechanical Engineers, 1989.

221. "Professionalism in Engineering Design: European Perspectives." *Proceedings of the Colloquium on Professionalism in Engineering Design.* London: Institution of Electrical Engineers, 1991.

222. Pugh, S. "Give the Designer a Chance: Can He Contribute to Hazard Reduction." *Product Liability International* 1, no. 9 (1979), pp. 223–225.

223. Pugh, S. "Load Lines: An Approach to Detail Design." *Production Engineer* 56, (1977), pp. 15–18.

224. Pugh, S. "Manufacturing Cost Information-The Needs of the Engineering Designer." *Proceedings of the Second International Conference Information for Designers.* Southampton: 1974.

225. Pugh, S. "Quality Assurance and Design: The Problem of Cost Versus Quality." *Quality Assurance* 4, (1978), pp. 3–6.

226. Pugh, S. "The Design Audit: How to Use It." *Proceedings of the Design Engineering Conference,* 1979, pp. 1–10.

227. Pugh, S. *Total Design: Integrated Methods for Successful Product Engineering.* Wokingham, England: Addison Wesley, 1991.

228. *Quality Assurance.* AMCP 702-3. Alexandria, VA: U.S. Army Material Command, 1968, pp. 5.1–5.9.

229. Ramamurti, V. *Computer Aided Design in Mechanical Engineering.* New Delhi: Tata McGraw-Hill, 1987.

230. Ray, M.S. *Elements of Engineering Design.* Englewood Cliffs, NJ: Prentice-Hall, 1985.

231. Ray, M.S.; and D.W. Johnston. *Chemical Engineering Design Project: A Case Study Approach.* New York: Gordon and Breach Science Publishers, 1989.

232. Red, W.E. *Engineering, The Career and the Profession.* Monterey, CA: Brooks/Cole, 1982.

233. Reddy, J.N.; C.S. Krishnamoorthy; and K.N. Seetharamu. *Finite Element Analysis for Engineering Design.* New York: Springer-Verlag, 1988.

234. Resnick, W. *Process Analysis and Design for Chemical Engineers.* New York: McGraw-Hill, 1981.

235. Rice, R.C.; B.N. Leis; and D. Nelson. *Fatigue Design Handbook.* Warrandale, PA: Society of Automotive Engineers, 1988.

236. Rich, M. *Knowledge-Based Engineering Gear Design.* Paper No. 90-773. Dearborn, MI: Society of Manufacturing Engineers, 1990.

237. Richards, L.G. *How Should We Teach Engineering Design?* Paper No. 83-13. Dearborn, MI: Society of Manufacturing Engineers, 1983.

238. "Design Theory and Methodology-DTM'90." ed. J.R. Rinderle. *Proceedings of the 2nd International Conference on Design Theory and Methodology.* New York: American Society of Mechanical Engineers, 1990.

239. Rosenbrock, H.H. "Designing Automated Systems-Need Skill Be Lost?" *Science and Public Policy* 10, no. 6 (1983), pp. 274–277.

240. Rumsey, H.A.; and R.L. Thompson. "Managing for Productivity in Engineering Design: Review and a Case Study." *Proceedings of the Second International Conference on Engineering Management.* Institute of Electrical and Electronic Engineers, 1989, pp. 89–94.

241. Rutter, P.A.; and A.S. Martin. *Management of Design Offices.* London: Telford, 1990.

242. Rychener, M.D. *Expert Systems for Engineering Design.* Boston: Academic Press, 1988.

243. *Handbook of Human Factors.* ed. G. Salvendy. New York: Wiley, 1987.

244. Sanders, M.S.; and E.J. McCormick. *Human Factors in Engineering and Design.* New York: McGraw-Hill, 1987.

245. Sandler, B. *Creative Machine Design.* Jamaica, NY: Paragon Press, 1985.

246. Sato, S.; H. Matsuhisa; and I. Sakamoto. "Decision Making in Design Process." *Memoirs of the Faculty of Engineering, Kyoto University* 51, (1989), pp. 218–234.

247. *Human Factors and Decision Making: Their Influence on Safety and Reliability.* ed. B.A. Sayers. London: Elsevier Applied Science, 1988.

248. Schapker, D.R. "Tort Reform and Design Professionals." *Journal of Professional Issues in Engineering* 116, (1990), pp. 258–265.

249. Schmid, C.F. *Statistical Graphics: Design Principles and Practices.* New York: Wiley, 1983.

250. Schmidtke, H. *Ergonomic Data for Equipment Design.* New York: Plenum Press, 1985.

251. Sherwin, K. *Engineering Design for Performance.* New York: Wiley, 1982.

252. Shigley, J.E.; and C.R. Mischke. *Mechanical Engineering Design.* New York: McGraw-Hill, 1989.

253. Shigley, J.E.; and L.D. Mitchell. *Mechanical Engineering Design.* New York: McGraw-Hill, 1983.

254. Siddall, J.N. *Analytical Decision-Making in Engineering Design.* Englewood Cliffs, NJ: Prentice-Hall, 1972.

255. Siddall, J.N. *Optimal Engineering Design: Principles and Applications.* New York: Marcel Dekker, 1982.

256. Siddall, J.N. *Probabilistic Engineering Design: Principles and Applications.* New York: Marcel Dekker, 1983.

257. Smith, R. *How to Plan, Design & Implement a Bad System.* New York: Petrocelli Books, 1981.

258. Spotts, M.F. *Design Engineering Projects.* Englewood Cliffs, NJ: Prentice-Hall, 1968.

259. Sruse, R.; and G.H. Powell. "Design Process Model for Computer Integrated Structural Engineering." *Engineering With Computers* 6, (1990), pp. 129–143.

260. "Statistical Quality Control in Manufacture Cannot Compensate for Poor Quality in Design." *Institute of Statisticians* 6, no. 2 (1987).

261. Stoll, H.W. *Simultaneous Engineering in the Conceptual Design Phase.* Paper No. 88-775. Dearborn, MI: Society of Manufacturing Engineers, 1988.

262. Stones, I. *Ergonomics: A Basic Guide.* Hamilton, Ontario, Canada: Canadian Centre for Occupational Health and Safety, 1989.

263. Suh, N.P. *The Principles of Design.* New York: Oxford University Press, 1990.

264. Swain, A.D. *Design Techniques for Improving Human Performance in Production.* London: Industrial and Commercial Techniques, Ltd., 1973, pp. 15–31.

265. Taguchi, G. *Introduction to Quality Engineering.* New York: Asian Productivity Organization, 1986.

266. Takeda, H.; S. Hamada; T. Tomiyama; and H. Yoshikawa. "Cognitive Approach to the Analysis of Design Processes." *Proceedings of the American Society of Mechanical Engineers 2nd International Conference on Design Theory and Methodology,* 1990, pp. 153–160.

267. Tang, J.C.; and L.J. Leifer. "Observations from an Empirical Study of the Workspace Activity of Design Teams." *Proceedings of the 1st International Conference on Design Theory and Methodology.* American Society of Mechanical Engineers, 1989, pp. 9–14.

268. Taoka, G.T. "Civil Engineering Design Professors Should Be Registered Engineers." *Journal of Professional Issues in Engineering* 115, (1989), pp. 235–240.

269. Tjavle, E. *A Short Course in Industrial Design.* London: Newnes-Butterworths, 1979.

270. Tomiyama, T. "Engineering Design Research in Japan." *Proceedings of the 2nd International Conference on Design Theory and Methodology,* 1990, pp. 219–223.

271. Trygg, L. "Engineering Design: Some Aspects of Product Development Efficiency." Ph.D. Thesis. Chalmers University of Technology, Goteborg, Sweden: 1991.

272. Ullman, D.G. *Mechanical Design Failure Analysis.* New York: Marcel Dekker, 1986.

273. Ullman, D.G. *The Mechanical Design Process.* New York: McGraw-Hill, 1992.

274. Ullman, D.G.; S. Wood, and D. Craig. "The Importance of Drawing in the Mechanical Design Process." *Computers and Graphics* 14, (1990), pp. 263–274.

275. Ulrich, G.D. *A Guide to Chemical Engineering Process Design and Economics.* New York: Wiley, 1984.

276. Vanderplaats, G.N. *Numerical Optimization Techniques for Engineering Design: With Applications.* New York: McGraw-Hill 1984.

277. Venable, W.S.; and R.K. Dean. "Structuring the Design Experience." *Proceedings of the Conference on Frontiers in Education.* New York: Institute of Electrical and Electronic Engineers, 1989, pp. 300–302.

278. Venturino, M. *Selected Readings in Human Factors.* Santa Monica, CA: Human Factors Society, 1990.

279. Vesilind, P.A. "Rules, Ethics, and Morals in Engineering Education." *Engineering Education.* February 1988, pp. 289–293.

280. Veyatyshev, V.F. "Structure of the Engineering Design Process and the Problem of Radio Engineer Training." *Telecommunication and Radio Engineering* 44, no. 4 (1989), pp. 101–106.

281. Vidosic, J.P. *Elements of Design Engineering.* New York: Wiley, 1969.

282. Wadden, R.A.; and P.A. Scheff. *Engineering Design for the Control of Workplace Hazards.* New York: McGraw-Hill, 1987.

283. Waldron, M.B.; K.J. Waldron; and K. Abdelhamied. "Differences in Reading Schematic Drawings of Mechanisms by Expert and Naive Mechanical Designers." *Proceedings of the 1st International Conference on Design Theory and Methodology,* 1989, pp. 15–21.

284. Wallace, K. *Engineering Design: Special Issue.* London: Butterworth & Co., 1983.

285. Walton, J. *Essentials of Engineering Design.* New York: West Publishing Company, 1991.

286. Walton, J.W. *Engineering Design.* New York: West Publishing Company, 1991.

287. Wang, Y.; and E. Sandgren. "New Dynamic Basis Algorithm for Solving Linear Programming Problems for Engineering Design." *Journal of Mechanisms, Transmissions, and Automation in Design* 112, (1990), pp. 208–214.

288. Ward, A.C. "Recursive Model for Managing the Design Process." *Proceedings of the 2nd American Society of Mechanical Engineers International Conference on Design Theory and Methodology,* 1990, pp. 47–52.

289. Weintraub, C. "Design for Manufacturability." *Design Graphics World,* October 1988, pp. 14–16.

290. West, D.; and V.L. Ferdinand. *A Management Guide to PERT/CPM.* Englewood Cliffs, NJ: Prentice-Hall, 1969.

291. Whitney, D.E. "Manufacturing by Design." *Harvard Business Review,* July-August 1988, pp. 83–91.

292. Wickens, C.D. *Engineering Psychology and Human Performance.* Columbus, OH: Merrill Publishing, 1984.

293. Wilcox, A.D. *Engineering Design: Project Guidelines.* Englewood Cliffs, NJ: Prentice-Hall, 1987.

294. Wilcox, A.D.; and L.P. Huelsman. *Engineering Design for Electrical Engineers.* Englewood Cliffs, NJ: Prentice-Hall, 1990.

295. Wilde, D.J. *Globally Optimum Design.* New York: Wiley, 1978.

296. Willem, R.A. "Design Science Interactions." *Proceedings of the 2nd American Society of Mechanical Engineering International Conference on Design Theory and Methodology,* 1990, pp. 323–325.

297. Wnuk, A.J. "Some Topics on a Model for the Design Process." *Proceedings of the Artificial Intelligence, Simulation and Planning in High Autonomy Systems Conference,* 1990, pp. 149–159.

298. Wood, W. "Martin Designers Zero In On Product Reliability." *Product Engineering* 37, (1966), pp. 100–101.

299. Woodson, T.T. *Introduction to Engineering Design.* New York: McGraw-Hill, 1966.

300. Woodson, W.E.; and D.W. Conover. *Human Engineering Guide for Equipment Designers.* Berkeley: University of California Press, 1965.

301. Woodson, W.E. *Human Factors Design Handbook.* New York: McGraw-Hill, 1981.

INDEX